Unboxing IT

Unboxing IT

A Look Inside the Information Technology Black Box

Christopher McCay

ROWMAN & LITTLEFIELD
Lanham • Boulder • New York • London

Published by Rowman & Littlefield
A wholly owned subsidiary of The Rowman & Littlefield Publishing Group, Inc.
4501 Forbes Boulevard, Suite 200, Lanham, Maryland 20706
www.rowman.com

Unit A, Whitacre Mews, 26-34 Stannary Street, London SE11 4AB

British Library Cataloguing in Publication Information Available

Library of Congress Cataloging-in-Publication Data

Names: McCay, Christopher.
Title: Unboxing IT : a look inside the information technology black box / Christopher McCay.
Description: Lanham : Rowman & Littlefield, [2016]
Identifiers: LCCN 2016004768 (print) | LCCN 2016008079 (ebook) | ISBN 9781475822113 (cloth : alk. paper) | ISBN 9781475822120 (pbk. : alk. paper) | ISBN 9781475822137 (Electronic)
Subjects: LCSH: Telecommunication—Popular works. | Computer networks—Popular works. | Information technology—Popular works.
Classification: LCC TK5101 .M377 2016 (print) | LCC TK5101 (ebook) | DDC 004—dc23 LC record available at http://lccn.loc.gov/2016004768

Printed in the United States of America

Contents

VI: Your Personal Box

VII: Closing the Box

Preface

Information technology can really scare the pants off of some people. It is complicated, nuanced, specialized, speaks its own language, and is truly invasive. It has crept into almost every aspect of our daily lives. It has changed the way we work. The way we learn. The things we buy. The *way* we buy. And it has become a driving force in the way the world is changing, connecting, and expanding.

While we have come a long way from thinking of everyone who understands IT as some kind of technowizard that dwells in a dark basement and makes strange things happen by waving his or her hands over a keyboard, to most people IT is still a complete mystery. A black box. I am astonished, almost daily, how little people truly understand about the technology we all use so frivolously. With this book I am hoping to demystify some of that world. To unbox IT.

Information technology is such a huge topic that to truly delve into every nook and cranny that exists within that world could take an entire volume of books. So think of this as an overview of a truly rich and dense topic of study. IT 101, as it were. I will attempt to give a basic understanding of the various disciplines that exist within the IT world, some of the thought processes behind the methodologies IT professionals follow, and touch on some of the realities of living in a truly technological wonderland.

We will look at computer hardware, software, networks, and much more. And while you certainly won't be ready to take on a vicious malware attack or design a fully virtualized disaster recovery site when you finish this book, you will have a greater understanding of what your friendly neighborhood IT tech goes through on a regular basis. (And also know what the blazes a "fully virtualized disaster recovery site" is!)

I feel this understanding is imperative for any nontechnology person—but especially for anyone in the education field. Technology advances at breakneck speeds, and understanding that world is a key to understanding the next generation. The current generation (my childrens' peer group) was born into a world with a computer in every classroom; they don't remember a time when we didn't all have access to the entire worldwide knowledge base in our pockets.

The base level of knowledge they possess is the equivalent of a decent entry-level help desk technician these days. I've been working in the information technology world for over fifteen years, and I have seen all kinds of crazy things. I started as one of those help desk technicians and have worked my way up to my current position as the head of IT for a national program management firm. I've designed networks from scratch, migrated entire companies from one backend server platform to an entirely new system (hardware and operating system), managed million-dollar IT budgets, and pulled many an all-nighter to ensure that a system update didn't disrupt the workday.

I have spent my career in the for-profit industry, but I have had the good fortune for the last five years to work for a company that is heavily involved in advancing the education world. We work with all levels, from K–12 to higher education. We work with individual charter schools, as well as entire school districts. I've also been married to a teacher for eighteen years, and we have two school-age children. I've seen the evolution of technology in the classroom, watched my children work a mouse and touchscreen without any practice or confusion, and, like all of us, I have lived through the evolution of technology in our lives.

My parents had to walk to the local library and sort through encyclopedias, old newspapers, and (if they were lucky) microfiche records of events. My children have access to the entire collected knowledge of the planet through a wire in our house. That is, for many of us, an unbelievable change in the way we view the world. And yeah, some days that weirds me out too.

Ultimately, my goal is to give you the tools to better understand this fully integrated aspect of our lives. Computers and technology have reached a level of invisibility in our lives that is equal to indoor plumbing and electricity. We don't think about the mechanics of how they work, but when any of those systems breaks it is a *massive* disruption in our lives. All of us understand the basics of how a fuse box works or how to plunge a toilet, and with this book you will learn those equivalents in the IT world.

SO WHAT?

So what is the point, really? My boss is very fond of that simple two-word question. And it really does rip right into the heart of everything, doesn't it? Why is all of this important? Why should we spend the next hundred pages or so together? And what can you hope to get out of this?

Well, if I haven't made it clear yet, technology is an omnipresent thing. We simply can't escape from it without completely cutting ourselves off from the world. And the people who have dedicated their lives to learning about IT, fixing IT, bettering IT, and presenting IT to everyone else are an eclectic, eccentric, sometimes frustrating, always fascinating breed of individuals. IT may be everywhere, but not everyone understands it or appreciates it. And more than a few people are downright terrified of it.

By the end of this book, I hope that you won't be one of those people. Instead, you will be someone who either works with, teaches, manages, or employs IT professionals, and you will have a much better understanding of what it is they do and what type of people they are.

Acknowledgments

While there are an almost countless number of people throughout my life that deserve thanks and praise, I will keep this fairly simple.

Mom and Dad, thanks for everything.

Nathaniel and Zander, you guys are the best boys any father could ask for. Sorry my deadline made me miss out on Halloween last year.

And Melissa. Thank you for everything. You are an amazing woman, and the best wife ever. I love you!

Introduction

As stated earlier, information technology is a huge topic to cover in one sitting. Obviously, there are multiyear advanced degrees on the subject. So what can we hope to accomplish with just one book? How about advancing from little to no knowledge to a basic entry-level understanding? Every IT professional agrees that everyone should have a base-level understanding of the technology world.

So many questions, frustrations, problems, and misunderstandings that IT professionals deal with on a daily basis could be solved by ensuring that everyone had this basic set of information. This is especially important in the ever-evolving world of education. Anyone who is school-age these days *does not know* what it is like to live in a world that does not hum with technology. And those who choose to work in the education field will be required to have a base-level technological understanding to properly interface with these kids.

This book will start with an overview of the history of information technology. It is a topic that, surprising to many of you, stretches back (at the writing of this book) almost two hundred years. Like all good stories, there are heroes and villains, wins and losses, and a glimpse into a magical future. And then the journey continues into the wonderful world of "talking tech." I'm sure you have noticed that the IT world speaks its own language. In fact, there are all kinds of subdialects within information technology.

Mostly, the "talking tech" chapter will concern itself with understanding some core technology terminology (a crash course, as it were) that will help you to understand many of the topics covered in the rest of the book. Don't worry; this won't be as dry as a dictionary. But just as you need to understand the symbols of arithmetic to have any kind of discussion about math, so too will you need this overview to understand IT.

After we have established our baseline we will dive into a discussion of IT as a function of the larger organization. Whatever the organization, be it a business, a school, a nonprofit, or a startup, they all have some set of requirements for what they want to get out of IT. What is often surprising is that those organizations aren't very good at knowing what it is they want from IT. Mostly this appears to be because the people in charge don't know enough about IT to know what they want. Many of them are operating from an old model that turns to technology for the absolute minimum set of business needs. It has long been a struggle within the IT world to try and change that. One of the problems, however, is that while organizations don't always know how to talk to IT, it is equally true that IT doesn't know how to talk to the organization.

Most IT professionals have progressed through a career track that has kept them elbow deep in technology. Then, one day, they are handed a title that includes the words *manager* or *supervisor* in it and the non-IT managers are surprised that the IT person has trouble adjusting from tech talk to business talk. It is an important goal to make sure that everyone who works in IT gets exposure to the nontechnical side of the businesses they are working for. We will be looking at both sides of the equation, then, to make sure that you can see what it is that both groups (information technology and the organization) need from each other.

How do you ensure that you are doing a good enough job at what you do? Do you track improvements? Are you reviewed annually? One of the things that most children dread in school is being graded. They never really understand the importance of tracking progress at the time. It is just what is expected. IT professionals, by comparison, *love* to be graded. Metrics are a key aspect of daily life in the IT world.

There is nothing more satisfying than an uptime dashboard that shows all systems green. Or a help desk satisfaction survey that shows you are meeting all of your service level agreements and your customer feedback has you rated at 97 percent. So you better believe we will spend some time covering important IT metrics and learning what they mean. We will also cover the metrics that are most important to the organization and to you as a technology user. Interestingly these are not always the same thing.

From there, we will venture into the wonderful world of hardware. Servers, laptops, desktops, tablets, smartphones; we will touch on all of those. There are so many aspects to the hardware we all use on a daily basis that no one really thinks about. Most people have heard the terms *hard drive, processor,* and *memory,* but do you understand them? (File cabinet, brain, and countertop by the way, but we will get into that later.) And the obvious place we will go from there is software.

Do you know why we need operating systems? Who knows just how many software processes are running when you are just staring at your desktop with *nothing* open? And we will look at the evolution of the app as we have become an increasingly mobile world. After those (for most people) relatively comfortable topics, we wander into one of the truly hidden areas of IT. We'll spend some time unpacking networking and understanding how information flows from one computer to another.

And no, it isn't magical fairies . . . it's spiders, of course. World Wide Web, and all that. Honestly, though, this is an area that even IT professionals find confusing, and anyone who is not an IT professional and can actually talk to their network technician about what he or she does will *greatly* impress them.

When you read the words *IT professional*, what is the first type of person that comes to mind? More than likely most of you are picturing someone wearing a headset, sitting in a cube, and answering questions about why something is broken. That would be the glamorous (insert your own sarcastic eye-roll here, please) life of the help desk technician. This is truly a skill set unto itself, and where most in IT started.

We will talk about the help desk process, why it is important that you do open a ticket properly instead of just asking for a favor, and cover some basic troubleshooting topics. And yes, before you ask, you should reboot. You'd be amazed what that can fix, and we will even talk about why that is. Oh, and save often. But you should know that one already.

That's it, right? We talked about IT as a part of an organization. We're going to look at what to measure, hardware, software, networks, and someone to fix things when it breaks. What else could there possibly be, *right*? You are picking up on the sarcasm in that last sentence, I hope. From these more visible areas of IT we will be shifting into the processes, methods, and secret sauce behind the scenes. Because there is an awful lot to this gig that non-IT folks don't ever really see.

Did you know that the IT profession has invented multiple methods for project management (PM)? Most people don't think of IT work as project-based, but so much of what happens inside the IT black box is project-driven: new system designs, network build-outs, software development, and so much more. These are the things that are going on at three in the morning so that the average worker doesn't see any downtime. And obviously if there wasn't a plan in place things would get ugly, and that's why we've designed these PM methodologies as flexible and redundant.

We will then discuss the never-ending cycle that is process improvement. Those who have chosen to make a living within the technology realm tend to be driven by the need for efficiency. How can we make it smaller? Faster? How can we take that manual process that you are doing that takes thirty minutes and automate it so it takes five? This is a fun, and frustrating, topic. And one that is becoming increasingly difficult in our current technical landscape.

With so much that we can do with just a few clicks, anything that takes more effort is seen as not good enough. And every time we raise the bar, we reset the expectation level. Hence the never-ending cycle.

Think we've covered everything yet? Come on. You know better than that! So let's talk risk. Risk management is another one of those "we can write a whole book on it" subjects. The simplest way to think about it is this: Where am I vulnerable, and what can we do to prevent that? There will always, *always*, be people out there who are attempting to use a lack of technology understanding on your part to their advantage.

It is unfortunate, and it says more about the human condition than we can cover here, but it is true. So what do we in IT do to prevent that? How do we prepare and plan? And we absolutely do prepare and plan! It would be astonishing to find that you haven't already met people who live for this battle.

No man is an island, and unless you work for a megacorporation, no IT department can do it alone. Vendors help us fill those gaps. Vendor management is an evolving area within IT. Who do the technology people turn to when they have questions about technology? Who supplies all that amazing technology? How do we choose who to purchase from? Who are our trusted consultants? And where is the line drawn as to which group is in charge of what service? And just who *is* answering when you call for support in the wee hours of the morning?

After all of that, there is still one more area of concern within information technology that we will need to cover. That is the topic of security. Security is another one of those broad topics that could easily fit into its own book, but it is an essential part of what we do in IT. Understanding where the risks are, who the risks are coming from, and how to attempt to mitigate those security risks is essential. And more and more, that understanding is important within our daily lives as well. This is, perhaps, the most universal topic to discuss given the high level of techno-logical integration into our daily lives.

All of that, in various levels of importance and depth of knowledge, is what is going on in the head of an IT professional. So let's talk about the kind of people that willingly take all of this on for their profession. There are so many different types of IT professionals, and we will examine an overview of all the different areas based on the information we will have already covered. From application developers to hardware engineers to help desk technicians, we will explore all of the kinds of folks who make their living in this crazy world.

We'll also dive into an understanding of how best to interface with IT professionals in ways that you, and they, will be able to learn from the experience. I have spent my entire career asking the question, "What can we learn here?" and I have found several scenarios where non-IT people can learn from their IT comrades and vice versa. We will talk about that, and even more importantly, when you need to get out of the way and let IT get the job done without having to stop and explain things.

It doesn't happen often (one would hope), but when the entire IT team is barreling through the hallway jabbering quickly to each other in what sounds like a combination of your high school algebra class and some weird foreign language (there is *a lot* of math in IT), remember to get out of the way! We promise; we'll explain as soon as we aren't worried about something crashing and causing some *very* expensive damage to the system.

And then we will round out all of these explorations of information technology with a dive into your personal digital life. As we are all well aware, there is no escaping technology anymore. Our refrigerators have Internet access now. Did you know that Internet-accessible appliances have already been hacked and used as bridging points for illegal spam-ming activity? So it's probably a good idea to put some thought into what all this tech means for everyone in their daily lives.

Additionally, we will look into what impact the growing consumer-ization of technology has had on IT in the workplace. The IT department doesn't hold all the keys to the technology kingdom anymore. Is that good or bad? Well, that depends on who you ask. And in the final area of exploration into our digital lives, we will look into the future. What will technology look like in just a few years?

And what can we expect as this generation, which has *never* known what it is to not be connected, comes of age? The current crop of technology explosions are the result of the generation that came of age with the dawn of the Internet; just imagine what will happen once the generation that could work a mouse *before* they could properly spell starts making their mark.

So make sure you have your work backed up. Reboot to ensure your memory is free. And let's get started.

I

The Dimensions of the Box

ONE

History of IT

How far back do you think the use of the computer goes? If you read the introduction, that is a bit of a trick question. The first machine to be considered a programmable computer was built by Charles Babbage. He designed what he termed an "analytical engine" that took input from a series of punch cards and could output the result of the programmed computation via a printer. His presentation of the project to Parliament apocryphally included this question from a member of Parliament: "If you input the wrong information, will it still output the answer you desire?" Thus proving from moment one that user error would be inherent in the usage of every computer from then on.

There were several computational machines that were designed and built in the first half of the twentieth century, but all of them were analog devices—meaning that they could only perform one computational task at a time. While this did increase the efficiency of the computation, it would not revolutionize public use of these systems. The next leap forward, from analog to digital, occurred when Alan Turing designed the first Turing machine. This was a computational device that utilized algorithmic mathematics. This device is widely held as the first "modern" computer.

Another key aspect of early computers was size. As you may or may not be aware, the physical devices required to perform the work were oversized and machined by hand. It wasn't until World War II that computers evolved into electromechanical devices. This allowed the process of miniaturization to begin. As more and more electronic elements were added to replace the large, cumbersome, moving physical parts, computer processors were able to shrink. Storage of data was still a problem at this time though, as most data was still being stored on magnetic tapes.

The invention of the transistor once more changed the landscape of computers. These smaller devices replaced the vacuum tubes that had been used up until that point. The final piece of the miniaturization puzzle snapped into place with the invention of the integrated circuit. This single device, made up of multiple transistors, is what continues to drive miniaturization today. As this chip gets smaller and faster, we can build more efficient portable devices.

PERSONAL COMPUTING

The more recent past is what is most interesting and exciting to the average computer user. In fact, the very *concept* of the average computer user only exists because of the personal computer (PC) revolution. Everyone has heard the stories of the scrappy kids in their garages who invented global brands. The mid-1970s into the early 1980s saw an explosion of home-built computers that would evolve into what we recognize today as a standard computer setup.

However, while the garage builders started a trend, it was IBM that finally brought PCs to the masses. For much of the 1980s PC hardware was either built of, or based on, IBM hardware. Apple, as we all know, was also prominent during this time. And the beginning of the Windows/Apple debate started here. Apple had brand-specific hardware and software. IBM, and the IBM clones that flooded the market, were all Windows-based.

Later, hardware and software will be covered in depth, but it is important to note some major historical landmarks that aided in the personal computer evolution. The earliest forms of computers required an input of commands in a programming format that the system recognized to tell the computer what to do. "Back in the day," as we older computer users like to say, input was done through punch cards. These were a stack of cards, roughly index card sized, with holes punched into them.

These holes, arranged in a format the card reader converted to internal commands, were the only way of feeding commands into the computer. And these early computers could only accept one computational input/output challenge at a time. As you can imagine, this did not make learning how to use and program a computer an easy thing. The addition of keyboard data entry was the first hurdle necessary to bringing computer use to the masses.

Additionally, the development of the operating system meant that computer users no longer had to tell the computer everything to do. The computer, through the OS, already had a set of internal commands and parameters. The addition of the mouse and the graphical user interface (GUI) completed the evolution of the early PC into a machine that was accessible to everyone.

A COMPUTER ON EVERY DESKTOP

The PC revolution meant a change in the way that businesses utilized computers. Until the advent of the PC, if a business made use of computers, it was in the form of a mainframe, which was a large centralized computer that was used to manage and calculate large data sets. It was not available to everyone, and to use it required specialized skill sets in computer operation and programming.

Time-sharing of computer resources began to evolve the use of the mainframe, as it opened up computer use and programming to a wider set of users. This concept, that everyone could benefit from the computational power of their own computer, helped fuel the PC revolution. By the mid-1980s, businesses had embraced the idea of the PC as a device that could enhance and enable the workforce. And by the turn of the century, most daily work tasks had been computerized to an ever-growing extent.

GOING MOBILE

As computer use increased and technology improved, the goal of a truly portable computer was achieved. Laptops have now become the standard computer form factor, and it is stunning to think how far we have come in so short a time. Mobile technology leapt forward again with the invention of the smartphone and the evolution of the tablet into a usable piece of mobile computing.

Along with mobile technology, the World Wide Web was invented. This system of interconnected devices has evolved alongside the technology over the last twenty years and has now become the primary way in which we interact with our mobile systems. We have developed a toolset that has enabled access to the collected knowledge of everyone that is connected to the Internet. Whether that is good or bad is a point that is debatable on its own. The mere fact of its existence, however, is astonishing.

COMPUTERS IN THE CLASSROOM

Since so many early computers were nothing more than giant mathematical analysis engines, it makes sense that many of their uses were for educational pursuits. While it wasn't until the PC revolution that most K–12 schools had access to computers of any kind, universities have played key roles in the development of computer invention and evolution. Turing did much of his early work while at Cambridge. Cambridge, along with the University of Pennsylvania and many others, continued to construct and refine various computer types.

With the advent of the PC, however, computer use as an educational tool became more ubiquitous. Early computer use focused on learning programming languages and teaching how to use the computer itself as a tool. However, once computer use itself became a skill set, utilizing computers to enhance the learning experience became the goal. And with the growth of technology in all of its various forms, computer-assisted learning has become the norm.

Computer-based learning and curriculums that are developed, presented, interacted with, and graded entirely on a computer are now a standard form of educational delivery. Schools at every level are using the Internet to enhance outreach and as a tool for developing new digital age skill sets with students. The current generation has had the benefit of growing up with computers as a standard tool. They inherently know and understand how to interact with them.

Additionally, the interactive capabilities that touch-screen technology provides allow students with multiple learning styles to learn in their own way. Educational software utilizes multiple sensory inputs and adaptive learning tools that adjust the difficulty levels based on the responses of the student. This allows the teacher to customize and individualize the curriculum in ways that can be challenging in a standard, non-computer-assisted classroom setting.

KEY IDEAS IN THIS CHAPTER

- Charles Babbage created the first analytical engine, which is considered the first computer.
- Alan Turing evolved the computer from analog to digital when he created the Turing machine.
- In World War II, computers evolved into electromechanical devices and began the process of miniaturization.
- The personal computing revolution exploded in the mid-1970s and led to the current environment of a computer in every home.
- The mid-1980s saw businesses embrace the PC as an enhancement device for the workforce.
- The Internet expansion and mobile technology has led to the current interconnected world that we exist in.
- Education has always been closely tied to computer evolution, as most early computer design and exploration was performed by universities.
- Computer-based learning has evolved to the point where we are no longer just learning *about* computers; they are a tool to assist with learning in general.

TWO
Talking Tech

IT has its own language, and to a nontechnological person it can sound very intimidating. As with all specialty areas, it has terms and phrases that are thrown around with ease by those in the know, and it is filled with acronyms and long strings of numbers. This internal language can make it very difficult to speak with IT professionals about what they do for any length of time. Many in the IT profession find themselves teaching about and explaining what it is they are attempting to do more than actually working on a problem. The sheer ubiquity with which technology has entered our daily lives can make it seem as though everyone *should* know more than they actually do about that technology, even though that is not the case.

This book is a very broad overview of the technology world. While we won't be diving deep into the weeds of how things work, it is important to be able to speak about and understand a common set of concepts within IT. This language barrier is usually what intimidates people the most about understanding IT. Once a technician starts talking about how the offsite disaster recovery site is made up of a virtualized environment that is updated across a high-speed full mesh topology to a facility with N plus 1 redundancy, that can all start to sound like the *wah-wah* noises the adults make in a Charlie Brown cartoon. So let's start small.

IT encompasses a wide variety of hardware and software types. Most of us are familiar with the "poster child" for IT, the computer. We are aware of the fact that a computer has a processor, memory, a keyboard, a hard drive, a motherboard, and a mouse. It is important to note that these basic pieces are replicated in most IT hardware. All technology, whether a server or a smartphone, has some version of those components. Obviously, the scale is different for each based on the size of the device.

It is also important to understand that any device that is Internet accessible has some form of NIC (network interface card) for communicating with the Internet. Addressing on the Internet is done by IP (Internet protocol) address. This is a string of numbers, much like a phone number, that identifies the location and physical connection port of the device.

IPv4 is the most prevalent form of this addressing, and it consists of four octets of up to three numbers each (8.8.8.8 or 67.215.71.202, for example). But because the human brain is pretty horrible at remembering long strings of numbers, we have developed a method for converting those numbers into names. DNS (domain name system) is the system that converts an IP address into a website or computer name. That way you only need to remember Google.com, not 74.125.226.7.

BINARY

At the fundamental level, all information on a computer comprises ones and zeroes. This is known as a binary counting system. Using an ever-increasing number of places, any number can be expressed in binary terms. 00 is obviously zero. 01 is one. 10, read in binary, is two. And the pattern continues in that manner. So it is possible to express a number like 137 in binary as a string of ones and zeroes like this: 10001001.

The reason for binary being used as the base counting system for computers is very simple. Electrical impulses can be either on (1) or off (0). So by stringing a series of on and off impulses together, any number can be generated. And by then converting those number strings into data formats (using various conversion software), the computer can read the binary input as the data we, the users, wish to see.

BITS AND BYTES

All data is stored electronically in a series of ones and zeroes. A computer's processor is designed to interpret strings of ones and zeroes and convert them into various characters, commands, inputs, and outputs. These strings are made up of bits and bytes. One bit is the smallest unit of data storage. It is either a one or a zero. Not much can be done with a single bit, so eight bits are grouped together to form a byte.

One byte is the standard unit for data storage. One byte is enough data to equal one typed letter in a word processing application. From there, we scale up metrically to the data storage sizes we are familiar with. A kilobyte (Kb) is a thousand bytes. A megabyte (Mb) is a million bytes, and most network data transmission happens in multiples of megabytes. Most hard drives come in multiples of gigabytes (Gb), which is a billion bytes.

However, we are increasingly seeing data storage mediums in multiples of terabytes (Tb), which is a trillion bytes. Storage mediums can, and do, go even higher than these numbers. Storage capacity now is limited only by the physical size and number of hard drives that can be interconnected. But the average user will only interact with storage sizes like these for the foreseeable future.

MOORE'S LAW

It is impossible to spend any significant amount of time talking about IT and the advancements within it without first discussing and understanding Moore's law. Gordon Moore, the cofounder of Intel, made an observation about computer chip capacity over time. He noted that capacity roughly doubled, while size and price stayed the same, every year to two years. And while this is not a universal law, like gravity, it has appeared to hold true over time.

Therefore, we can assume that the ability for a computer to perform advanced functions will continue to increase all while size and cost will lower over time. This is one of the driving forces behind the miniaturization and mobilization movements. This exponential curve has yet to level out. And while there have been bumps and slowdowns, on average the curve has continued to rise.

USERS

Without a doubt, you have heard an IT professional refer to anyone who is *not* a member of the IT team as a "user." Some people, I am sure, find this term slightly jarring and possibly derogatory. It is not, however, meant in that way. It simply means those who use the computers, as opposed to administrators (the IT department) who oversee the computers.

And while it is true that, from time to time, IT professionals can get frustrated with non-IT folks, the use of the general term is meant only as shorthand and not as an insult. Therefore, from time to time, the term will be used in that way throughout this book.

EDUCATIONAL TECHNOLOGY

The world has come a long way when it comes to IT. We have invented, customized, streamlined, and miniaturized the components that make up the computer so rapidly (thanks in part to the effects of Moore's law) that within three generations we have gone from a select number of highly skilled users putting men on the moon to virtually *everyone* having a computer in their pocket that has more computing power than the Apollo 11.

The education world has been hard pressed to keep up with these rapid fire advancements in technology. When I was in school, we had one small computer lab with eight black and white televisions hooked up to basic PCs. Now, my children have multiple computers per classroom and laptops at home. And there are still a number of teachers in the workforce who were there when I was in school. So how can it *not* be hard for these advances to impact methodologies and curriculums?

The debate over technology and whether it is being used too much or not enough is not the subject of this book. No matter where you fall on that particular subject, it is still important to understand how to communicate with a generation that is eternally plugged in, and how to teach anyone, of any age at this point, how to navigate in a world that is becoming increasingly technology and Internet dependent.

As such, it is important to understand why it is that technological advancements seem to be happening at a faster and faster rate (once again, Moore's law rears its head) and how to adjust to and anticipate the changing needs of the classroom. We have already seen, with even a minimal amount of technological advancement, that the traditional classroom structure can be exploded using remote connection tools and educational software platforms.

It is no longer necessary to be in the same room, or even the same continent, as the students you are teaching. And while this creates its own set of educational challenges, the capacity for educational advancement through these methods can't be diminished. Distance and correspondence learning are not new concepts, but the interactivity that technology allows is what has changed the game.

The challenge you will now face comes in taking this overall understanding of the IT world presented in the following pages and adapting it to your own environment. There will be a few specific educational technologies that we discuss, but the thrust of this work is to gain an overall understanding of IT as a whole, so we won't be spending too much time on specific educational tools.

KEY IDEAS IN THIS CHAPTER

- To better understand the IT world, there are some basic terms and concepts that need to be understood.
- All computer-based technology consists of the same basic parts: processor, memory, keyboard, hard drive, mouse, motherboard, and NIC.
- Addressing on the Internet is done by IP (Internet protocol) address.
- All information on a computer comprises ones and zeroes. This is known as a binary counting system.
- One bit is the smallest unit of data storage. It is either a one or a zero.
- Eight bits are grouped together to form a byte. Large amounts of bytes make up data storage rates.
- Moore's law states that capacity roughly doubles, while size and price stay the same, every year to two years.

II

What's In the Box

THREE

What Do the Organization and IT Want from Each Other?

IT really is everywhere. Nobody thinks about how integrated into our lives and our workplaces it has become. Until, of course, it stops working properly. A very busy workday can come crashing down by something as simple as an Internet connection not properly working. So with that in mind, what exactly does an organization want from the IT department?

Well, the basics would seem pretty straightforward: a working Internet connection, e-mail, a platform for creating work product, and minimal interruptions in the daily work flow. That last one, while it sounds very simple, is what keeps many a systems and network engineer up at night—both metaphorically *and* literally, as major changes and upgrades need to happen during off hours.

Obviously, the reality of what the organization needs from IT is much deeper and complicated than that. But this basic framework is helpful to keep in mind. Technology needs to work properly. Technology needs to enhance the workflow of the day-to-day, not get in the way of getting work done, and, perhaps most importantly, make that day-to-day work easier. This means that a good IT group is constantly striving to find the better, faster, more efficient way to accomplish these goals.

How do these goals translate to the educational world? It's all well and good to talk about IT as a function of business, but the same basic framework can be applied at all levels of education. The primary difference is that the day-to-day work isn't a report or a product; it is teaching. And technology has become deeply integrated into the classroom. So much so that we now have classrooms that are entirely virtual! They exist *only* because of technology.

EDUCATIONAL NEEDS

The modern classroom houses many IT devices. Whether it is an elementary class or a graduate-level one, computers are everywhere. All students will interact with some type of computer multiple times within any given week. That could be for something as simple as typing up a paper, or as interactive as a website-based learning program. Computers have become an essential teaching tool.

IT is also integrated into the communication process between teachers, students, and parents. E-mail is the medium in which information is disseminated to large groups by the school administration. Websites listing classroom assignments, rules, and information are the norm. Universities use social media to promote themselves, stay in touch with alumni, and recruit new students. Rapid response information is sent via text messages when necessary.

Students today have an innate understanding of the technology world. Multitasking with multiple devices comes easy to them. Therefore, utilizing various technology tools in a classroom setting can help to enhance the learning environment. YouTube allows for demonstrations of activities with relative ease. Music of all varieties and styles is available to sample and learn from. Cultural information and examples are at our fingertips for historical and current event analysis.

Where technology in the classroom can truly excel, though, is in enabling the various individual learning styles of the students. Touch sensitive technology allows for more hands-on learning for those who respond better to that. Translation technology is available to assist with non-native-speaking students and for better teaching tools in language study. Audio/video options are almost too numerous to count and are available via simple Internet access. And learning through play and story software allows for even the youngest students to find ways to interact with and learn from technology.

These needs are the same no matter the grade level, no matter the location. However, the ability for certain school districts to fulfill those needs can be difficult. Some rural school districts can have more problems accessing reliable high-speed Internet connections. Technology advancements happen at a rate that is hard for many school districts, no matter the size or location, to keep up with.

Additionally, educators at the K–12 level tend to feel that their primary use of technology is for administrative tasks. While they are able to use some technology as a teaching enhancement, the access for every student to benefit from the technology is limited. This changes dramatically at the college level, as most schools now make it a requirement that all students have their own computers, thus enabling the use of technology as a teaching tool more readily.

Therefore, what the elementary classroom needs most from IT is the understanding that what little technological means are available need to stretch as far as possible. This has long been one of the areas where IT excels the most: finding those areas where efficiency and productivity meet and making more happen with less.

While it is true that budgeting and availability issues certainly come into play with this particular need, many state and local agencies are working on making technology more available to as many students as possible. Especially at the local level, school districts are aware of the differentiator that technology can be. And as technology continues to evolve and become more affordable, more schools will be able to make IT available to more students.

The technology industry as a whole is also making strides in learning what education needs from IT. Mobile technology, custom learning apps, and education-specific hardware are being turned out by the IT world on a regular basis. The need for IT-based learning, which has been fueled by the current generation's constant need for newer and better ways to input data, has already led to numerous learning websites and "brain game" apps.

Special education can benefit from IT in multiple ways as well. Assistive and adaptive technology has grown by leaps and bounds as more emphasis is being placed on those arenas. Physical disabilities are no longer the hurdle to education they once were with the assistance of various technologies designed specifically to aid those with various physical limitations. Computers that can read eye movement, reading applications that translate words on a page into spoken language, hardware that can translate websites into Braille—all of these are just a few examples of the ways in which IT can be used to enable and enhance the learning environment.

What education as a whole needs from technology is not that different from what the organization needs: enhancement and efficiency, nontraditional methodology that enables change and growth, a means to connect with and manage the goals of those who are using the technology, and possibly some radical changes to approaches that aren't working the same way they used to.

BUSINESS NEEDS

Modern business is deeply tied to IT. There is no aspect of the business world that does not utilize technology in some way to perform daily tasks and functions. No matter what your product is, technology is helping to make it happen. This is true whether your end product is a technology itself, or if your end product is nothing more than consultation and analysis.

Where business needs differ most from educational needs are that the end product of education is teaching, which can be assisted by and sometimes even performed entirely by technology, while the end product of business is whatever that business is selling, and more often than not that end product can *only* be achieved through the use of technology. Reports and analysis require software, and computers on which to do them. Assembly lines are automated and computer monitored for efficiency. Construction is managed and monitored through various technology tools and actually achieved through massive applications of technology and hardware.

Technology is a requirement of modern business, not an option. Therefore, business requires IT to be available and aware of what makes the business function. IT must find where the problem areas are in the daily operation of a business and work to determine if new or different IT solutions can solve those problems. Business needs IT to know and anticipate what new technologies can make a business achieve its goals better.

Most importantly, business needs IT to be unobtrusive. IT is a tool, and repeatedly we have talked about how important efficiency and enhancement are. However, IT can also be highly disruptive. New technology and changes to what we are accustomed to can drastically reduce the ability of a business to function if not properly prepared and rolled out. It is essential for IT to ensure that all essential functions of the business are maintained without causing massive delays.

IT NEEDS

As much as we may all wish we were, IT professionals are not miracle workers. Yes, we have been known to make amazing things happen with little time and even less money. We are all experts at finding creative ways to solve the problems that are presented to us. But even we have our limitations. And usually those limitations are defined by one of three things: time, money, or the current limits of technology.

The most important thing that the organization can understand about IT is what is and what is not possible. Believe it or not, there is an outer limit to what is possible via technology. That limit is constantly moving, and Moore's law shows that the limit expands very rapidly, but it is still there nonetheless. Most businesses are not reliant on the bleeding edge of technology. That can be a scary place to work, as brand-new and experimental IT can't be relied on for the kind of execution consistency that is required to ensure business profitability.

Some businesses and educational facilities are designed specifically to live on that outer edge. They are aware of the risks involved in using experimental technology, and they have prepared themselves with contingencies and backup options for when the inevitable breakdown occurs. Also, they usually have enhanced amounts of the other two dynamics of the limitation equation (time and money), which means they are not set back drastically when a bleeding-edge piece of IT fails to perform properly.

That is a key understanding that IT needs the organization to have. Any one of the three aspects can be pushed to the limit to enhance the IT experience. But if that happens, then the other two aspects *have* to be maintained and respected. IT can find a solution to any problem quickly, but it will be expensive and can't rely on untried technology. If the solution has to cost as little as possible, then it will take time and once again can't rely on untested technology. And as we've already discussed, if testing the untested is the goal, then there will need to be a lot of time and money to recover if and when there are problems.

COMMUNICATION IS KEY

As with all good relationships, IT and the organization need to know how to communicate with each other. In the next chapter, we will delve into this much more, but for now, let it be enough to know that open communication between the two is essential. IT is a function of the greater whole and can only create the necessary enhancements and efficiencies when it is treated as such. While IT has become as operationally expected as water and electricity, there is much more to the requirements IT has to fulfill than either of those utilities.

KEY IDEAS IN THIS CHAPTER

- The organization needs technology to work properly, to enhance the workflow of the day-to-day, to not get in the way of getting work done, and, perhaps most importantly, to be able to make that day-to-day work easier.
- Education needs technology for administrative as well as instructional use.
- Where business needs differ most from educational needs are that the end product of education is teaching, which can be assisted by and sometimes even performed entirely by technology, while the end product of business is whatever that business is selling, and more often than not that end product can *only* be achieved through the use of technology.
- The most important thing that an organization can understand about IT is what is and what is not possible.
- IT limitations are usually defined by one of three things: time, money, or the current limits of technology.

FOUR
Communicating with the Organization

One of the hardest things for an IT professional to do is learn how to *not* talk tech with anyone who asks them about IT. The goal of the "Talking Tech" chapter was to ensure that everyone had a solid baseline understanding of certain terms and conditions within the IT world that we who work within it simply take for granted. And it can be difficult for an IT professional to remember that not everyone functions with that same baseline of knowledge.

As a result, communication between IT and the organization can become strained and difficult sometimes. IT professionals can be curt and dry when communicating. We like to deal in crisp realities. We do, after all, work in the world of binary. Something either is or is not. And that is the lens through which many IT professionals view the world. The organization, however, functions within a world where technology can be viewed as a utility.

Therefore, it is easy to misunderstand how much effort goes into ensuring that systems and technology are functioning properly at all times. This communication gap can be very difficult to overcome, but is essential for the proper functioning of IT within any given organization. Let us explore this gap further and find ways to overcome it on both sides of the equation.

HOW IT SHOULD APPROACH COMMUNICATING WITH THE ORGANIZATION

The first thing any and all IT personnel need to keep in mind is that, unless they work for a technology firm, the organization is not as deeply engrained in technology as they are. Any and all acronyms and techno-speak must be explained thoroughly. In fact, it is best to attempt to avoid them at all costs. Sometimes it is impossible to explain why a malfunction has occurred without getting technical. When that is necessary, ensure that all technical terms are explained within the sentence they are used in.

Analogies are an IT professional's best friend when it comes to communication. The phrase "kind of like this" is one that is used often. While not many people understand what a full mesh is, they do understand what a spider web looks like. The network chapter of this book relies heavily on picturing network traffic as a road, not a water pipe. The more grounded and ubiquitous the reference, the more people will be able to understand the concept that is being explained.

It is also important for an IT professional to remember that although metrics (as we will discover in the next chapter) are a favorite tool for IT, most people have very little interest in discussing the five 9s of uptime or how low the emergency change numbers are (99.999 percent system availability and infrequent emergency system updates). They *do*, though, want to know that, yes, the server is available for saving their file and there's nothing for them to worry about with the equipment.

To an IT person, those are far too simplistic ways of looking at those problems. But to non-IT people, any more information is simply too much. They have their own job-related technical language, and they simply don't have room in their personal hard drive (brain, to all you non-tech folks) for the IT speak. And the more the IT professional uses terminology that is hard for them to understand, the less they will want to even approach the IT team when they have questions or problems.

The one time all of this gets thrown out the window, and an overuse of technical terminology is required, is when explaining an unexpected outage or some kind of emergency problem fix. Once the problem has been resolved, an extensive outage explanation should be circulated to the entire organization. Yes, the technical terms will need to be explained within the report, but it is essential to ensuring the organization maintains trust and confidence in the ability of the IT department to handle these kinds of problems. When something is broken, that's when the organization wants to know the IT team is capable of handling any problem.

As with any professional communication, it is important, especially when discussing some form of emergency that needed to be resolved, to maintain the proper tone. The organization can get nervous when technology problems happen, as it becomes so apparent in those moments just how reliant on technology we have all become. Like a doctor with a patient, the IT team must maintain a level of calm and ensure that all explanations are understood by those who were affected by the problem. Patience and thoroughness are the key.

COMMUNICATING WITH IT

Yes, IT professionals can be overly technical sometimes when we talk, but we are capable of communicating like human beings. It can be valuable, much as we discussed earlier, to gain some baseline knowledge of the IT world first—that is, after all, the goal of this book. But even if you know nothing about IT, it is still possible to effectively communicate with the IT department without becoming completely overwhelmed.

Firstly, it is important to know that IT professionals love to solve problems. It is, in fact, one of the few things that are universal among all of the various IT professions. We can't *stand* it when there is a problem in front of us and we aren't able to find a solution. It will eat away at an IT professional until the solution is found! But that doesn't mean that we are always available to help solve your particular problem the second you walk in the door.

The largest pet peeve of any IT professional is that most people who approach the IT department assume that there is nothing going on at the moment. When there is no emergency going on, no outage being tended to, and no flurry of activity by the IT department over some apparent issue, that does *not* mean that there is no work being done. In fact, one of the most frustrating truths about the IT world is that often work means waiting.

Much like cooking, sometimes you just have to let it sit. An IT professional can, in fact, be working on multiple machines at any given time and if one is downloading updates, another is loading software, and yet another is running an internal diagnostic test—and that is a lot to juggle.

But it looks like the IT pro is simply sitting and doing nothing. That is why it can be very frustrating when someone approaches who doesn't understand that this is what IT work looks like sometimes and assumes the IT pro has a lot of availability to work on their problem.

So while you do not need to approach IT like they are a group of caged animals who may strike out at any moment, you also should not assume they are a group of bored teenagers who just feel like staring at the wall. Yes, IT will be happy to help you with a problem, but please don't be offended if they need a moment or two before they can be available to work on it. And then don't be surprised if, like a dog with a bone, they simply don't give up until your problem is completely solved.

When bringing your problems to IT, make sure you have as much information as possible. We will look into the wonderful world of the help desk a little later and cover several troubleshooting steps that you can try yourself. But details of a problem are always the best way to get a problem solved. And having those details already with you when coming to IT about the issue is the best way to get a solution started quickly.

While the first inclination of anyone is to assume that the only thing the IT department can fix is your computer or a broken application, one of the things that IT excels at is process improvement. We will look into the various methods and processes IT uses for this later, but it is important to know that we are interested in making your work better and easier to accomplish. IT doesn't implement a new system simply because it would be fun (usually); they do it to make an improvement on the way things are being done. To find a better way.

Consulting IT on upcoming projects or issues with current workflows can only be beneficial to all parties. The earlier IT can be involved in something that may need a technology solution, the better. Remember the three factors discussed last chapter. More time means there is more wiggle room when it comes to budget and the kinds of technology needed for the solution. Even if the initial thought for the project does not include technology, involving IT can result in an unexpected solution.

IT is one of the few areas that has insight into almost all areas and levels of the organization. IT is aware of which other departments may be facing the same types of problems and can help them interact to find solutions. IT is aware of newer methodologies that are being deployed by certain personnel because they needed IT to help get things set up. IT is capable of bridging the gaps in knowledge across the organization, even if they don't have the knowledge themselves. IT knows where to point everyone to get the knowledge.

MAINTAINING COMMUNICATION

It is important for both IT and the organization to remember that open communication is beneficial to all. IT houses and maintains essential, and sometimes confidential, information for the organization. As such, IT can sometimes become insular and wary of too much communication. Yes, security is important, but isolationism is not. And the organization needs to remember that IT is not a group of antisocial misfits that are unable to talk about anything other than technology, although we will be happy to discuss the cool new tech toys if you'd like.

KEY IDEAS IN THIS CHAPTER

- One of the hardest things for an IT professional to do is learn how to *not* talk tech with anyone who asks them about IT.
- The first thing any and all IT personnel need to keep in mind is that, unless they work for a technology firm, the organization is not as deeply engrained in technology as they are.
- Analogies are an IT professional's best friend when it comes to communication.
- It can be valuable to gain some baseline knowledge of the IT world to effectively communicate with the IT department without becoming completely overwhelmed.
- The largest pet peeve of any IT professional is that most people who approach the IT department assume that there is nothing going on at the moment.
- When bringing your problems to IT, make sure you have as much information as possible about the problem.
- The earlier IT can be involved in something that may need a technology solution, the better.

FIVE

What Numbers to Watch

It is widely known that what is measured is what is understood. It is also true that what is measured is what gets acted upon. Metrics are a key aspect of information technology. So much of what we do in the IT world can be impacted by bad system performance, so it is vital to track and understand what the systems are doing at all times. It is also important to be able to report out usage, availability, return on investment, and more to ensure the continued viability of IT within an organization.

IT METRICS

We will start by looking at a set of standard internal IT metrics. These are the things that matter to the IT professional. They allow the IT department to understand their environment, make any necessary changes, and gauge the need for updates and upgrades. They can also be used as a troubleshooting device when there are problems. Without understanding what *normal* looks like, it can be difficult to respond to *abnormal* properly.

The first, and some would argue most important, metric is uptime. Uptime measures how long a given system has remained powered on. This is essential to understanding possible wear and tear on the physical parts of a system. It is also essential to ensuring that whatever task that system is performing is being performed. Various systems require different levels of uptime. Network equipment, for example, requires near 100 percent uptime to ensure that access to all internal and Internet-based resources are available. Printers, however, can have windows of scheduled downtime that will have minimal impact on workload.

Side by side with uptime is availability. There is a key distinction between uptime and availability in that a machine can be powered on, but if it is not properly network accessible, it is not available. So uptime may not be impacted, but availability will be. Some will measure these two things as one metric, but as you can see, there is a slight but key distinction between them. And that distinction is very important, because it can be highly frustrating to have the response to "I can't get to something" be "Well, it's turned on."

Once system availability and uptime is being recorded, it is important to measure any and all impacts on system usage. Incident metrics are used to measure problems that have occurred. Incident metrics usually track several different things: overall time to resolve individual incidents, average time to resolve incidents over a period of time, number of incidents within a given period of time, and number of users affected by an incident.

While it is the goal of every IT department to have zero problems with their systems, it is inevitable that something will occur. How that is dealt with and recorded for future learning and efficiency is vitally important. As such, incident metrics can prove to be the most important and telling indicator about the overall IT system infrastructure. Keeping them minimized is important, but most important is keeping the time to resolution numbers as small as possible. Most people understand that problems occur, but extended periods of downtime will lead to complaints.

System performance during normal usage is also important to measure. How the various pieces of the infrastructure behave when not in a crisis is essential to ensuring that you can maintain (and return to) a normal state. It is also good to remember that while data within a given system is electronic, the hardware itself still consists of moving parts that can break down over time. Usage metrics help to anticipate the need for repairs and replacement of devices.

Application metrics look an awful lot like system metrics. Enterprise level applications also need to maintain uptime and availability. They require incident management and usage metrics. They also require metrics that monitor their interaction with the various systems and other applications. Are the applications utilizing the hardware systems efficiently? Do updates to the software require updates to the hardware?

Does the application suite function and interact properly with other applications? These are all key things to measure and track.

Changes are also important to measure and track. Change management metrics track planned and emergency changes to hardware and software. Change success rates are important to understanding how often attempted changes run into issues. And total number of required changes is important to understanding the overall stability of the infrastructure. Once changes have been made, the quality of the system or application performance is tracked for a period of time to measure the overall quality of the change.

ORGANIZATION METRICS

Many of the metrics that should be tracked and analyzed, as we have seen, are essential to the IT department understanding their world. Most non-IT executives are not interested in any of what we have just covered. However, there are a number of metrics that are essential to the operation of an IT department that are important for the organization to understand. These will have a direct impact on budgets, staffing levels, and overall satisfaction with the IT department.

While there are many important things to measure, the essentials for reporting IT performance and viability to the organization can be broken down into three key areas: employee satisfaction, security status, and financial status. There are numerous aspects to each of these areas, but we will cover why this breakdown cuts directly to the heart of the organization understanding IT's role within it.

Several metrics that have already been mentioned as IT-specific metrics can be rolled up into an overall business-level metric that shows infrastructure stability. Metrics for uptime and availability, change management, and incident management can be rolled up into an overall metric for stability. Most non-IT folks aren't interested in the minutiae of how and why the system is functioning. They just want to know that it is, it will continue to, and changes won't impact employee performance. This in turn is one piece that will roll up into the employee satisfaction metric.

Staff satisfaction is measured both internally to IT, and with every other employee's satisfaction with IT performance. Internal IT staff satisfaction and performance is a key management metric. It is also important to rate and measure the usage of external vendor support. Help desk metrics will include such things as number of tickets over a given period, average time to close a ticket, number of tickets resolved within service level agreements (SLAs), and overall satisfaction with the help desk experience. All of these will roll up into a satisfaction metric that in turn will also be a piece of the employee satisfaction metric.

Security incidents are an unfortunately high-profile issue within the IT world. Because of the growing reliance on technology to perform our daily work tasks, and the fact that so much sensitive data is now being housed by so many areas within an organization, it is essential that security be measured and reported on at the organization level. Number and types of security incidents must be measured and tracked. Risk exposure, potential loss of data, and damage to contractual obligations (both to clients and to staff) must also be measured and tracked.

Financial metrics that should be measured may seem simple (how much does all this technology cost?) but can be fairly complex. While budget usage is an important metric to track, it is not the be all and end all of the financial matrix within IT. Return on investment and cost mediation are key aspects of reporting on how technology usage is lowering the cost of doing business. For example, how does the cost of a video communication system offset the benefit of no longer needing to travel to visit every client?

Additionally, tracking application and system implementation cost against the time regained by the staff now that a given application or system is faster and more efficient can show the impact IT makes on the overall financial bottom line. IT departments that generate product directly for a client have their own set of financial goals and metrics to measure. But this book is focusing on IT as a supporting aspect and not a revenue generation center.

EDUCATION METRICS

While the idea of metrics defining educational needs or changes can be hotly debated, it is important to note there are IT-based metrics that can be used within education that are more specifically geared for that environment. Several of the examples mentioned above are easily transferable, as system stability and user satisfaction with those systems is fairly universal. Security is an area that is even more imperative in educational

IT. Family and child information, safety, and security are all highly regulated areas and as such must be monitored and maintained.

It is absolutely true that attempting to use metrics to drive any changes or evaluations of the current educational landscape would be hotly debated and criticized. However, as more and more businesses and organizations are looking to utilize data analysis and trends to anticipate changes and needs within their respective environments, the same tools could be used to help anticipate needs in the educational arena.

Enrollment trends, demographic changes over time, overall response to any changes in curriculum: these are just a few examples of metrics that, if tracked over a specified period, could be used to gauge what kinds of adjustments specific schools, school districts, or entire state infrastructures may need to make. Remember, what gets measured can be made better!

As the debate over the educational viability of certain types of technology use continues, metrics will become essential to determining the outcome. The addition of technology-based curriculum will be measured and monitored. The variances between traditional brick-and-mortar-based education and entirely digital-based education must be measured. The different ways that individual students react and respond to different delivery methods is also essential information.

It should also be noted that these metrics, while in their infancy, will change drastically very quickly. The current generation, and all subsequent generations to come, have been born into a world where technology is the status quo. They expect and respond to a different kind of interactivity with technology. They are not awed or intimidated by IT. For them, it is as impressive as an oven or a toaster. Therefore, the expectations for how technology is woven into the educational experience will grow exponentially, and we must be ready to anticipate that. And only that which is measured can be prepared for.

KEY IDEAS IN THIS CHAPTER

- Metrics are a key aspect of information technology.
- They allow the IT department to understand their environment, make any necessary changes, and gauge the need for updates and upgrades.

- They can also be used as a troubleshooting device when there are problems. Without understanding what *normal* looks like, it can be difficult to respond to *abnormal* properly.
- Some key IT metrics are uptime, availability, incident resolution, system performance, and change management.
- IT performance and viability to the organization can be broken down into three key areas: employee satisfaction, security status, and financial status.
- As more and more businesses and organizations are looking to utilize data analysis and trends to anticipate changes and needs within their respective environments, the same tools could be used to help anticipate needs in the educational arena.

III

The Walls of the Box

SIX
Hardware

Everyone knows what a computer is, I'm sure. However, as we have already discussed, computers are really nothing more than processing engines. And those engines can be housed in all kinds of different physical devices. For the next several pages, we will talk about all of the various types of hardware we can call "computers." Additionally, we will talk about the hardware that runs the background infrastructure nobody outside of IT ever thinks about.

When is the last time you wondered how the information gets from the ether that is "the Internet" to your laptop as you are browsing? Or how the phone call you are taking on that smartphone actually knows it's a phone call and not a text? The hidden world of networking is something we know we will be discussing in depth later on, but here we will be learning about the various devices that power that part of the IT world.

COMPUTER HARDWARE BASICS

A computer is any device with a processor, a hard drive for storage, and memory to allow the various programs to run. Your car has a computer in it. So does your air conditioner, your water heater, your cell phone, and possibly even your refrigerator. But what most people understand and recognize as a computer falls into one of two standard form factors: a desktop or a laptop.

Desktop computers, as we know from our overview of IT history, became commonplace in the 1980s. They were large tower boxes that would connect to (in most cases) an old television for a monitor. Desktop computers stole the keyboard input from old typewriters, and once the GUI (graphical user interface) came around, allowing for a mouse and pointer on the screen, we had what we would all recognize as a standard desktop setup. Once the GUI became so essential, monitors themselves began to evolve to give the best possible visual experience.

The inside of the tower is where all of the magic happens. There the computer is broken down into various parts to manage the various tasks. The motherboard is what houses all of the connections and what moves electricity throughout the device. This is where the serial bus is located, and it has modular slots for expansion and addition of hardware.

The most common modular components that are added to enhance the user experience are graphics cards, sound cards, and network cards. Graphics cards are designed specifically to enhance the visual experience. They have their own independent processor and memory, so they aren't drawing on resources that are being shared by other parts of the computer. Sound cards work the same way, to enhance the audio experience. We will cover network connection hardware shortly.

The processor is the brain of the computer. This is where all of the computation occurs. And, at its heart, it is still Babbage's computational device. It opens and closes various gates to perform input and output procedures, and in a series of on and off sequences (performed in binary) that happen in incalculable fractions of a second, it makes decisions and changes to the software.

The hard drive is the storage for the device. This is where all of the software is kept for the processor to process, along with the operating system (OS). This is also where all data is stored for later retrieval. And without this the computer would be nothing but a simple (albeit fast) addition/subtraction machine. I refer to this as the "file cabinet" of the system. It is where everything is kept. Which is *very* different from the memory of the computer.

Memory is the working space of the computer. The processor, as powerful as it is, has a limited amount of space in which it can hold information. The hard drive, as large as it is, is not designed for quickly feeding information to the processor and then pulling out the responses from the processor. The memory is designed to do just that. When a program needs to run, it actively moves the information that it needs to the memory. The memory then feeds into, and retrieves from, the processor. And then returns the finished program data to the hard drive for permanent storage.

These three key components work together to form the basic functionality of the computer. And it is easy to understand now why the specifications for these three devices are used to sell the computer. A faster processor allows for faster and more complicated "thinking." A bigger hard drive allows for more data storage, more advanced program use (due to the space the program information takes up), and longer useful life of the machine. More memory allows for faster program speeds, multiple programs to run efficiently at the same time, and smoother overall operational feel.

In addition to these devices, computers also have network connection hardware. Generally this is a hardwire connection port (RJ-45), but increasingly this is a wireless network connection device. Wireless connections are essential for laptop usage, and more often will be found in desktop hardware. It is important to note that wireless network connectivity is *not* as reliable or as fast as a wired connection. Those who have grown used to lightning fast Internet connection can find this fact frustrating when they are unable to connect via a hardline.

Computers also require some form of input device. The standard set of inputs is a keyboard and mouse. The mouse, as we know, is used to control the cursor and pointer on the screen to interact with the various pictograms that make up the GUI. And the keyboard is used to input characters. Increasingly, we are seeing a touch-screen interface to allow direct screen contact to take the place of the mouse and cursor. And there are other forms of input such as eye-tracking and voice activation.

Laptop computers, as the name implies, are portable versions of computers. They contain the same internal devices as can be found in a desktop computer. And, as we move further up the exponential curve of Moore's law, the hardware has shrunk enough and become fast enough to make laptops a viable alternative for most people. The days when a desktop computer was, by definition, best to have are long behind us. In fact, most desktop devices can now rival a server-sized machine in capabilities, which has made these larger versions of computers also less necessary.

Servers are the big brothers of the desktop. Servers are designed to run far more complex and resource-intensive systems, such as the enterprise-level components that run an organization. E-mail systems, file systems, websites, and more are run from these bigger and beefier versions of the standard computer. They usually include multiple processors, much higher levels of memory, and terabytes of data storage. These enhanced hardware components allow them to be faster and respond to more usage requests than the average computer.

STANDARD NETWORK EQUIPMENT

Networking equipment falls into three basic categories: switch, router, firewall. Many of us are most familiar with the term *firewall*, as we have some passing understanding that this helps to protect us from computer viruses. But the other devices that move information and power across networks are much more elusive.

The network cable that snakes out of the back of the computer and connects into the wall jack is connected back to a central network closet. At this location, all of the various cables are terminated into a patch panel. This panel is a static connection from the network closet to the wall jack. The network cable that attaches from your computer to the wall jack is a modular component. There is another cable that connects the patch panel to the switch. The switch is the first step in the networking system.

Switches gather network connections and facilitate the movement of information from one device to another. They regulate power and data flow. In the networking chapter we will go into more detail about how this is accomplished. But the easiest way to think of them is as telephone operators. They take an incoming request for one device to communicate with another and send that request along.

Routers are the brains of the networking world. Routers know where all of the devices actually physically connect to the switch. Switches ask the routers to help move traffic around when the data request comes from a device that is not directly connected to the switch. Routers are the power horses of the networking world, and they make the Internet flow.

Firewalls are, as the name implies, the walls that protect and guard individual networks. Networks are broken down into internal and external connections. Anything that talks directly to the Internet is externally facing. Anything that does not is internally facing and needs to traverse through the firewall to access information on the Internet. Firewalls are used to control what data is allowed into a network, which devices are allowed to talk into and out of the network, and what should be blocked.

WIRELESS DEVICES

Wireless devices are a special kind of router. They perform the function of a router, without having to be physically connected to the devices that talk to it. They maintain the same kind of routing information internally, but use over-the-air communication rather than direct wire communication.

BASIC PHONE HARDWARE

Increasingly phones are becoming integrated parts of the IT infrastructure. As voice over IP (VoIP) connectivity increases in usage, we are seeing phones use more of the same types of hardware that the computer infrastructure uses. Phones now connect to the same types of switches as computers do. They need routers to differentiate the types of traffic from data and voice. And they require their own types of firewall connections due to the type of information that is being passed out to the Internet.

Handsets have come a long way from the days of Ma Bell. Now we have Internet-ready devices for our phones. These handsets require a standard RJ-45 network connection. They have their own internal computers that maintain the firmware that allows the device to function. They can be modular and upgradable, much like a standard computer. And increasingly they use some form of visual interface in addition to the standard number touchpad.

The phone infrastructure is run by a PBX (private branch exchange). This is a centralized system, usually a computer running specialized software, that maintains all of the various routing information for the phones. It knows what phone numbers the phones are allowed to claim as their own and how to communicate out to the larger telecom world to connect to phones that are not part of the internal system. This PBX is the telecom version of a data router.

PORTABLE DEVICES

What makes a tablet different from a laptop? Most would say it is the interface type. Tablets are built and designed for a touch-screen interface. A keyboard and mouse *can* be used, but they aren't required for interfacing with the device. Additionally, the internal hardware is not as robust. The processor, hard drive, and memory are smaller and lighter. This increases the portability factor but decreases the types of complex programs that can be run on these devices. That is why more complex software, like video and photo editing, won't work on a tablet.

What makes a smartphone different from a tablet? Again, it is a matter of size and intended purpose. A smartphone, at least in original intent, was built to be a phone first and a computer second. Now, however, it seems as though they are designed the other way around, with the phone features being secondary. But they do still exist and this sets the smartphone apart (albeit slightly) from the tablet.

As we now enter the era of wearable devices (smart glasses, smart watches, etc.), we will find that our definition of what is a "computer" becomes more and more blurred. As these devices become smaller, lighter, and more portable, they are no longer built to resemble a desktop at all. Input devices are now designed for all of our senses (eye motion, voice control, body movement, etc.) rather than just the standard keyboard and mouse. And as we move increasingly up the right-hand side of the Moore's law curve, we will find these devices become more and more portable, changing the form factor virtually overnight. But they are all, still, computers.

KEY IDEAS IN THIS CHAPTER

- A computer is any device with a processor, a hard drive for storage, and memory to allow the various programs to run.
- The motherboard is what houses all of the connections and what moves electricity throughout the device.
- The processor is the brain of the computer.
- The hard drive is the storage for the device.
- Memory is the working space of the computer.
- In addition to these devices, computers also have network connection hardware and some form of input device.

- Networking equipment falls into three basic categories: switch, router, and firewall.
- Switches gather network connections and facilitate the movement of information from one device to another.
- Switches ask the routers to help move traffic around when the data request comes from a device that is not directly connected to the switch.
- Firewalls are the walls that protect and guard individual networks.
- Wireless devices are a special kind of router. They perform the function of a router without having to be physically connected to the devices that talk to it.
- Phones now connect to the same types of switches as computers do.

SEVEN
Software

With our basic understanding of various types of hardware complete, we can now turn to understanding the software that allows these devices to run. As was discussed earlier, without the various types of software that have been developed over the years, a computer would still be nothing more than a highly advanced abacus. An *extremely* efficient mathematical engine, but not anything more than that. Software is where the magic happens.

Software is what allows us to type on a keyboard and have words appear on the screen. It allows us to use a mouse to maneuver through files and application lists rather than having to input long and tedious strings of commands to tell the computer what we want it to do. And increasingly, it is what allows us to simply touch the screen and literally put a world of knowledge at our fingertips.

OPERATING SYSTEMS

Any discussion of software has to start at the most basic level. What runs the basics of the computer? We are all familiar with the term *operating system*, but I doubt we have put a lot of thought into what that system does. At any given moment, the operating system has dozens of hidden functions running to present to the user the view we have all become accustomed to when we look at a computer screen. Most of us are used to seeing a background of some kind (usually with a picture we have chosen ourselves) with a series of icons on it. This graphical interface is called a GUI (graphical user interface) and has been the standard for personal computers for over thirty years.

However, this interface was created for the ease of use of the human user and not for computational efficiency. As was discussed in the historical overview of IT, user input has gone through several iterations and upgrades over the centuries. The GUI was the final hurdle needed to bring computer use to the masses. No longer were long strings of commands necessary to tell the computer what you wanted accomplished. Everything was available at the click, or double-click, of a mouse.

But why is an operating system important? If computers are simply devices that are designed to do what we tell them to, why do we need any kind of framing system that takes up resources and space on a computer? Why don't we simply have applications that live directly on the hardware? Wouldn't that make things faster? The short answer is no.

Operating systems are essential to telling all of the various parts and pieces of a computer how to function. Without an OS, the various hardware pieces would be just that: pieces. They would have no way of knowing what tasks to perform, or what order to perform those tasks in. As an operator, you would have to individually control each piece of the machine and tell it how to manipulate the application that you are attempting to run. The OS is your hidden helper, overseeing and managing all of the thousands of tasks that are required in the proper sequence to ensure that your computer does what you ask it to do.

This means that even when you as a user of the computer aren't doing anything, the computer itself is doing lots of things. As soon as you turn on the computer, the OS is controlling and manipulating hardware. It is determining what personal settings you have saved for how you like to view and interact with the machine. It is learning and confirming what hardware is available to it and what purposes those hardware pieces are designed for. Has the hardware state changed since last start-up? If so, what new processes will be needed to manage that new hardware? What application software is available, and what resources will those applications need? All of these are management tasks that the OS is performing constantly, without you even being aware.

FIRMWARE

Personal computers aren't the only devices that require an OS to function. Any computer-assisted device needs a set of operational parameters. On most other devices we call this the firmware. It is the basic set of operational instructions that allow the device to know what its hardware is, what its task list is, and how to perform the basic operations that are expected of it. Your phone has firmware. Your car has firmware. And every piece of networking equipment that is used to enable the Internet to function has firmware. It is, at its core, a base-level operating system, but it is coded and manipulated very differently.

Most casual technology users will never need to worry about how to access firmware. However, this may change as our world evolves into a more computer-driven society. You may, someday, be required to run firmware diagnostics and upgrades on your personal appliances. Your car may need annual updates. And the more that this basic operational system is understood as a necessary part of system function and maintenance, the more you will understand about the way things work in the world around you.

APPLICATIONS

So what is an application? We should all be able to answer this question. It is, at its most basic level, a set of commands that tell the computer to perform a series of tasks. Whether that be as simple as typing in a document editor, or as complicated as running a space shuttle launch, it still comes down to executing a specific set of tasks in the proper order. But how does that work?

Software design has become highly complicated. That is driven, mostly, by our needs for any given application to be able to do hundreds of things at once. We don't want to just type. We want to control type size, font style, color, placement, and countless other style choices. We want the computer to anticipate what we *meant* to type, so it will tell us that something was misspelled or grammatically incorrect.

We want to be able to move large quantities of information across several different kinds of applications, which means that those applications need to be able to understand and communicate with each other.

And that is just a small example of the complications involved in what we, as users of technology, have asked software to do for us.

This is why software can no longer be stored on a small floppy disk and must instead be downloaded directly over the Internet. It is also why hard drive sizes have needed to grow so large for individual machines, and why it is still possible, with all of the amazing power and speed that modern computers have available to them, for a computer to eventually run slowly. Application software is highly specialized and takes up a decent amount of physical space on your hard drive even when it is not in active use.

When a given piece of software is in active use, no matter what kind of software it is, it goes through a standard set of hardware interactions. First, it is read by the OS from the hard drive. The application commands are then moved to the processor for, wait for it, processing. Most processors, as we have discussed, have a limited amount of storage space for housing commands and data that a given piece of software will require to function.

So the OS will load the required commands, data, and functions that can't live on the processor into the memory. This process, moving resources from hard drive to processor to memory and back again, is repeated for each additional piece of software that you have open. And some applications are more resource intensive than others. This is why processor speed, memory size, and hard drive read/write speeds are so important to determining what kinds of software you will be able to use.

This process is also why it is possible to overload the system. Remember, data at its most basic level is housed as a series of binary ones and zeroes within the computer hardware. And this data is moved and manipulated via electrical impulses that travel throughout the motherboard. The more applications that are open and running, the more the hardware is taxed. And, as we all know, this can lead to slower performance as more and more data is required to be manipulated.

This is also how corruption can sometimes occur. When data is being moved and manipulated, sometimes the OS misplaces a one or a zero. This causes data corruption. And this can lead to broken applications, lost work, or a computer crash. Additionally, it is possible to overload the amount of data that is housed within active memory. This can also lead to a systems crash.

SOFTWARE DEVELOPMENT

Because of the intricacies involved in the interaction applications have with the physical equipment it runs on, software development has become a highly specialized skill. It can differ widely based on the types of applications being designed, the hardware it is being designed for, or the OS it is being designed to operate under. Software development has its own long, storied history, and many of the innovations created for the process of designing and building software have led to the various types of project management methodologies used today. We will be delving deeper into those methodologies later in the book.

The problems involved in dealing with so many potential hardware configurations when developing and designing software can't possibly be stressed enough. Each computer processor has its own internal method for inputting and outputting data, known as the architecture of the processor. Each additional hardware component that the processor needs to speak to when the OS or software requires it also has its own internal architecture.

Because computer hardware is not standardized completely (there are "families" of different hardware types that allow for some levels of standardization), software can't simply be designed one way for universal use across all platforms. For a piece of software to operate on different brands of machine, especially when those machines have different operating systems, the software must be coded differently for the specific types of different hardware and OS combinations.

This is why, when certain types of hardware become too old, they can no longer run updated software. That hardware type is no longer being supported by the software code. This is also why it is important to note what type of OS the software you wish to buy functions on, and whether or not the new piece of hardware you are purchasing is compatible with the OS and software that you currently own.

DESIGN OVERVIEW

As with most creative efforts, software development begins with an idea, a function that is desired. Whether for a game, a work-enhancement product, a new video streaming technology, or the latest mobile app for "life hacking," all software begins with the initial idea of what is desired as the end product. How to get from that initial idea to the end product, however, is a long and involved process. Generally software development involves a team of programmers and designers who are working on various pieces of the software puzzle.

Software is designed and coded with modular pieces for multiple reasons. It is faster and more efficient to create it that way. It also allows the finished product to load only the modular functions that it requires to perform an action into memory and the processor. Also, if there are functions that will need to be repeated multiple times in multiple areas of the code, breaking those out into one-time functions that can be called when needed helps to keep the code clean and efficient.

There are too many coding languages to name here, but it is important to note that programming is a language skill. Certain programming languages are used for different types of application development. But all of them are built around a logical framework that consists of a series of commands that tell the hardware of the computer what to do and when to do it. Those commands can also be self-referential as they tell the application how to deal with various parts of its own internal logic.

Software coding languages allow us to interact with and manage the tasks of the computer without having to talk to the computer directly in its own internal language. As stated before, computers only understand binary (ones and zeroes) because the hardware is able to interpret the electrical impulses as either on or off. Each coding language exists to help manage and manipulate those ones and zeroes in different ways.

WEB DESIGN

Website software is its own specialized subset of software. Since the Internet has become such a ubiquitous part of our daily lives, it is easy to gloss over and forget how complicated this system truly is. There are numerous browser software available for accessing the information on the Internet. These various browsers are built and coded slightly differently. This is why some websites will work better in different browsers and why some website functions will only work in certain browser types.

It is also important to note that website design is its own specialty area as well. Website design was one of the areas that really brought the overall look and interactive experience of the process of using software to the forefront. Most software that existed before the Internet was based on more of a functional-needs format. The functions and options that the user was more likely to want most often were presented first, and so on.

Website design, and subsequent growth of mobile applications, has revolutionized the way that software is thought of. Now, how the person who is using the software can move around the screen and access the various functions of the software is *far* more prevalent in the designer's thoughts. User experience is its own job function and area within software development. And customer feedback on how easy it is to use a given piece of software drives changes more often than advancements in hardware nowadays.

MOBILE APPLICATIONS

Mobile applications (apps) are the most recent iteration in software design. They are smaller, more compact pieces of software that are built specifically for mobile hardware platforms. The explosion in the mobile hardware marketplace has forced software development to make another evolution. Now, software can no longer be hardware-specific. The sheer volume of different hardware styles in the mobile world have required software developers to create multiple iterations of their own software so that it can function on multiple hardware platforms.

This movement toward hardware agnosticism has opened up the world of software development to more areas of growth than ever before. Now a development team needs to know how the end goal required function will perform on any and all available (and future) pieces of hardware. This will lead to even more interactive capabilities between software types, and even more opportunities for the software development world to evolve.

KEY IDEAS IN THIS CHAPTER

- Without the various types of software that have been developed over the years, a computer would still be nothing more than a highly advanced abacus.
- Operating systems are essential to telling all of the various parts and pieces of a computer how to function.
- An application is a set of commands that tell the computer to perform a series of tasks.

- When a given piece of software is in active use, no matter what kind of software it is, it goes through a standard set of hardware interactions. Some applications are more resource intensive than others.
- This process is also why it is possible to overload the system and how corruption can sometimes occur.
- Because of the intricacies involved in the interaction that applications have with the physical equipment they run on, software development has become a highly specialized skill.
- Software is designed and coded with modular pieces to keep the code as clean and efficient as possible.
- Various Internet browsers are built and coded slightly differently. This is why some websites will work better in different browsers, and why some website functions will only work in certain browser types.
- Thanks to the increase in mobile hardware and consumerization, software can no longer be hardware-specific.

EIGHT

Networks

How information moves across a network can seem almost magical. Most people tend to think of it like water flowing through a pipe. This analogy is most likely perpetuated by the fact that we tend to refer to the data transmission medium as "the pipe." This is, unfortunately, not a very useful analogy in helping us understand how and why data moves across the network in certain ways. A better way of thinking about it would be a road, one with very specific rules for the order and speed in which cars are allowed to move.

NETWORK TRAFFIC

As with everything else at the root of IT, it starts with ones and zeroes. The data that is being transmitted is packaged within a data frame, and that frame is then packaged within a network packet. The network packet includes addressing information and allows for error detection. The layering of this information is called the OSI (open systems interconnection) stack. This model has seven layers and becomes increasingly complex as you move up the stack from layer 1 to layer 7. We will only be concerning ourselves with a deeper understanding of the first three layers.

In most standard Internet-based network connections, the first three layers will determine how traffic is packaged, presented, and routed throughout the network. Addressing, sending, and receiving all occur in these three layers. We will briefly discuss some of the functions of the higher layers, but they are not as important to a base level understanding of networking. Networking, perhaps even more than coding, is a complex rabbit hole that can confuse even the most seasoned IT professional.

Layer 1 is the physical layer. This is where the raw transmission occurs. This layer is defined by the medium in which the data is being transmitted. Whether it is cable or radio waves, how the data moves has a direct effect on transmission speed. This is also, in basic network troubleshooting, the first place you look to fix a problem. Is the cable still plugged in? Has it gone bad? Is there something that could be causing RF (radio frequency) interference and disrupting your wireless signal? All of these are possible layer 1 transmission problems.

Layer 2 is the data link layer. This provides data transfer between two connected nodes. There is some error detection that can occur at this layer within the network, and flow control is also possible here. However, since layer 2 is only concerned with point-to-point data transfer, it is limited. If the endpoint that you are trying to contact is not directly connected, then transmission can't occur. While this isn't necessarily an issue within a room, this *severely* limits the number of connections that can occur over any kind of distance.

Layer 2 devices are only aware of the other devices they are directly connected to. They maintain this listing within a MAC (media access control) table. All network connection devices have a unique MAC address. When a device is attempting to create a point-to-point connection, it will search for the MAC address of the other device by sending an ARP (address resolution protocol) request. This request and response process is how a layer 2 device learns which devices are directly reachable.

Layer 3 is the network layer. This layer is where routing occurs. Routes are used to reach devices that are not directly connected to your device. A routing table allows traffic to request a certain physical destination across the network. A router establishes a translation between the physical connection ports it has, and which network addresses can be reached using those ports. Routers are essential in network traffic flowing properly over any kind of distance, from as short a distance as between floors in a building to as long a distance as across oceans.

ANATOMY OF A PACKET

With this very basic understanding of network traffic flow established, it is important to dissect what is in a network packet. This is how network traffic shaping occurs. Different packet types will be treated differently and are ordered based on established protocols. This is why the water pipe analogy doesn't work properly. Water through a pipe simply flows, or it doesn't. It doesn't line up according to molecule type or number. And thinking of the Internet as just a big flow of data without understanding *how* that data moves can cause frustration when something doesn't behave the way you would expect.

Different types of protocols use different types of packets. So to attempt to avoid any more deep technological overload, we will be looking at a standard Internet data packet. It includes network addressing information (from both where it originated and where it is going), error detection components, a length indicator, a prioritization indicator, and the actual data. That is, as you can see, a lot going on. There are a few things it is very important to note.

The network addressing information is critical to ensuring that data is transmitted to the correct place. As we saw above, if the network location the data is trying to reach is not immediately connected, then routing is required. The desired endpoint address will determine how the router sends the data on to the next location. Depending on where that location is, it could travel through a series of routers. Each time it enters a new router, the packet needs to be read and processed forward.

Another key aspect, and the one that makes the water pipe analogy truly break down, is the prioritization indicator. Quality of service (QoS) is a system by which a router can be told what *type* of traffic should be allowed to travel across the network more frequently. Where we see this most often is with voice and video traffic. If voice and video are not given proper priority, and are forced to travel through the network along with other data mixing into the transmission, then video jumping and voice quality disruptions occur.

By giving these types of network traffic a higher priority, more of them are allowed to travel together. This keeps the quality of the phone call, or the video stream, intact. When you are on a VoIP phone and the person on the other end starts to clip and drop out, this is the reason why. This is also the reason why simply increasing the size of your network connection does not always fix these kinds of problems. If you are simply letting in more overall traffic, it doesn't fix the order in which that traffic is traveling.

NETWORK DEVICES

As was discussed in the hardware chapter, there are three basic network devices in the standard Internet capable network: a switch, a router, and a firewall. Thus far, we have touched on the layers of network traffic protocol and mentioned routing and what that is. Now we will take a deeper look at these different types of devices.

If data transmission is a road, then switches are the on and off ramps. They are the first, and most prolific, network device within network management. Switches do the bulk of the work directing network traffic, only passing off to the router the packets they are unable to manage themselves. Switches are usually layer 2 devices; however, it is possible to implement layer 3 capable switches. This minimizes the amount of hardware required on a local network, as the one device can handle both the switching and routing traffic.

Routers are the traffic directors of the networking world. They know where everything should be going, and through QoS they know the order in which everything should proceed. As we learned in chapter 2, "Talking Tech," IP addressing is the method by which devices track physical address location on the Internet. Route tables are made up of blocks of IP addresses that are translated to physical ports on the router. Thus, when a request for a specific IP address comes to the router, it is able to determine which physical wire to send that request on to.

Firewalls are the gatekeepers of the network realm. They allow traffic to flow into and out of private networks based on a set of rules established by the network administrator. These rules are known as ACLs (access control lists). An ACL determines what types of traffic are allowed to pass and can be programmed to allow only certain protocols to have access. Here, some layer 4 to 7 interaction happens as the transport protocols on layer 4 can be used to determine proper network access types.

Firewalls also act as the main route out from private networks to the Internet at large. As such, they can also be programmed to ensure that only certain types of traffic are allowed *out* to ensure that no harmful problems that may occur within your private network affect the Internet, if possible. As an example, a server that is designed to transmit website traffic can have all other types of traffic (e-mail, file transfer, etc.) locked down at the firewall level. This limits the exposure to the Internet at large and minimizes the risk of that machine being used for malicious intent.

WIRELESS CONNECTIVITY

Wireless routers are a very different, and yet very similar, type of device. They perform all of the functions of a router, but also operate a radio frequency antenna that broadcasts the availability of network connectivity within a given range. This broadcast capability can be hidden, but most wireless routers choose a password encryption method to manage user access. Their range is only limited by the signal power of their antennae.

Wireless, by the very definition of its transmission medium, is a slower method for moving data. Most standard wireless is capable of moving data at speeds up to 150 Mb/S (megabytes per second). This is fairly decent; however, most wired connections now move at speeds of multiple Gb/S (gigabytes per second), which is exponentially faster. Additionally, wireless signals can degrade due to environmental interference. Other RF signals, such as remote controls and cell phones, can cause transmission interference. Physical objects can also cause issues by either warping or blocking the transmission signal.

INTERNET BROWSING

For most people, the network is equal to the Internet. As we can now see, the network is in fact much larger than that. It encompasses all connected devices, whether they are hosting a website or not. However, Internet browsing is our most frequent interaction with anything that moves data across the various network devices and mediums that we have been discussing. Websites are coded using layer 7 applications that can be read by browser applications. We are familiar with most of them; Internet Explorer and Chrome are a few examples.

How a web browser functions is also a fairly complex program to understand, but again there are important pieces of the software it is good to be aware of. Browsers use different encoding and decoding technology, which is why different browsers can render websites in different ways. The primary set of website development tools, though, is readable by all standard browsers. Websites are created, and read into the browser, in zones. This is why pieces of a website may appear before others if you are on a slower connection.

And this is—it is important to note—how some viruses are able to enter a system. It is possible for a virus to be injected into a computer through malicious code on a website. As the various zones are loaded into a web browser, the code pieces have to move through the processor and potentially into memory for interactive system use. Malicious code can be written in such a way that once it is read into the processor it activates itself as a separate program.

Website antivirus and ad blocking software is designed to anticipate this possibility and prevent it from occurring. We will be discussing viruses and how to deal with them in more depth in chapter 14.

KEY IDEAS IN THIS CHAPTER

- Most people tend to think of network traffic like water flowing through a pipe. But a better way of thinking about it would be as a road, one with very specific rules for the order and speed in which cars are allowed to move.
- In most standard Internet-based network connections, the first three layers of the OSI (open systems interconnection) stack will determine how traffic is packaged, presented, and routed throughout the network.
- Network traffic shaping occurs when different network packet types are routed and treated differently based on established protocols.
- Quality of service (QoS) is a system by which a router can be told what *type* of traffic should be allowed to travel across the network more frequently.
- Switches do the bulk of the work directing network traffic, only passing off to the router the packets they are unable to manage themselves.
- Routers are the traffic directors of the networking world. They know where everything should be going, and through QoS they know the order in which everything should proceed.

- Firewalls are the gatekeepers of the network realm. They allow traffic to flow into and out of private networks based on a set of rules, known as ACLs (access control lists), established by the network administrator.
- Wireless, by the very definition of its transmission medium, is a slower method for moving data.
- Browsers use different encoding and decoding technology, which is why different browsers can render websites in different ways.
- It is possible for a virus to be injected into a computer through malicious code on a website. Website antivirus and ad blocking software is designed to anticipate this possibility and prevent it from occurring.

NINE
Help Desk

While we will be looking into the various job positions and functions in the IT world later, it is important to understand the various aspects of the help desk a little more deeply. Everyone who works in the IT field has had to assist with troubleshooting to some degree. And no matter how high you are promoted, there will always be someone asking for help with some aspect of their technology. As we become more personally reliant on technology, it is important to understand what the requirements of the help desk are and what you can do beforehand to help the process along to a rapid solution.

Anyone who has owned any kind of technological device has dealt with problems that they could not fix themselves. The help desk exists, obviously, for these occasions. Most help desks are set up along the same lines. There is an entry level at which you provide detailed information about what the problem is. Some systems are designed to take this information and find the appropriate technician to respond. Others have a generalist technician who will attempt to fix the problem first before escalating to a specialist.

The escalation process is one that can be frustrating but is essential to the efficient operation of the overall help desk. The end of that statement is important to note. It is for the efficiency of the *overall* help desk, not necessarily for the efficiency of your particular problem. No tiered system is designed specifically to make the resolution of your problem take longer or delay the process, but the queuing and tiered systems are in place so that the specialists can focus on the harder problems in the order in which they are presented.

Within many school districts, this tiered system is broken down along the generalist/specialist framework, but it is also broken down geographically as well. Many schools have a school-based tech specialist. This person is there to respond to very basic IT needs within one or two local schools. However, their abilities and access levels may be limited. They are the first line of defense and are there for quick turnaround requests.

The next tier up is an area technician. This person is the one who responds to all server level and infrastructure needs, as well as all of the more complex problems that the tier 1 specialist is unable to accomplish. Any requests that move to this level, by definition, will take longer to resolve. This extra time is important to note and be prepared for. And above the tier 2 IT personnel there are the district-level IT teams. They are responsible for all of the backend technology that keeps everything running (e-mail, websites, network topologies, etc.).

While at the college level, we find that most IT help desks are staffed by students of the university. Again, there is a tiered system in place with the higher levels staffed by full-time employees of the university. But no matter what type of help desk your particular educational organization uses, it is important to know and understand how to navigate and communicate within it.

TICKETING SYSTEMS

While this may sound obvious, when confronted with wait times and what seem to be obstacles to the resolution of your particular problem, many people lose patience with this type of system very quickly. They do not understand why it is important to take the time to create a tracking ticket for what they believe to be a simple and fast problem fix or question. But the tracking system is in place for multiple reasons.

It is there so that if more than one technician is required to work on a given problem, there are notes on what has already been done—thus saving you from having to go over the same troubleshooting steps every time someone new works on the problem. It is also there so that the technicians themselves can track resolution options for when these types of problems occur in the future, thus building the efficiency of future problem resolution.

Additionally, the ticketing system is there to ensure that work is completed to your satisfaction. It gives you a reference point for what has or has not been accomplished. And it allows you, as the person who started the process, to request more input and assistance until a proper resolution is found. While it will not help the process to beleaguer and hound the technicians, you can request that a ticket not be listed as closed until you are satisfied. This will keep the problem an active presence in the technician's workload.

SERVICE LEVEL AGREEMENTS

Service level agreements (SLAs) are an essential part of the help desk. They establish the contract that the help desk makes to the employees of the organization. SLAs determine the amount of time it will take to respond to a given issue. Each issue is categorized by the level of the SLA that is falls under. Generally SLAs are listed as low, medium, high, and system down. And each level has a corresponding severity level and response time anticipated. It is important to note that SLAs guarantee *response* time, and not resolution time. Some SLAs can include remediation clauses for any actions that will take longer than anticipated, but it is impossible to guarantee resolution time without knowing precisely what the problem is.

PERSONAL TROUBLESHOOTING

There are several things that you can do before initiating a help desk ticket to make the process go faster and smoother for everyone. It is important to know what kind of machine you have, what your operating system is, and as much detail about the problem you are experiencing as possible. Your machine make and model should be listed, along with some kind of identifying serial number on the outer casing. Most tablets and smartphones have a general information area in the settings that has this information.

Your OS can be found in the main computer settings. It is important to note what version number of the OS you are using, and what update level as well. Along with this information, any error codes that you have experienced will greatly increase IT's ability to find a rapid resolution to your problem. Lacking any error codes, the details of what you were doing or attempting to do when you experienced the problem can be very helpful. How many other applications were running? Did you hear any sounds when the problem occurred? Details like this can help to hone in on the problem.

There are also several troubleshooting steps that you can take yourself before attempting to contact the help desk. The one that is joked about most often is rebooting. "Have you turned it off and back on?" is a question that is asked almost every time a help desk ticket is opened. While this seems like a lazy way to fix a problem, there are very valid reasons why it can be helpful. As we discussed in the software chapter, all applications are just moving ones and zeroes around from the hard drive to the processor and the memory. And that is where corruption and problems can sometimes occur.

Rebooting the machine causes the memory buffer to be released. It empties the active programs out of the processor and resets the OS to the initial startup configuration. This can clear out many issues at once that without a reboot could require extensive troubleshooting to find the exact problem location. A reboot also forces the OS to once again query the various pieces of hardware that are attached to the system to determine what is and isn't available as a resource. This can clear up hardware usage issues and make connections function properly.

If a reboot does not fix your immediate issue, then physical problems should be the next thing that you look into. Are all of the cables securely connected? Is everything plugged in and powered on? Are there power issues anywhere else in your area that could be causing issues with your system? Does your work area have sufficient space around your machine to allow air flow for cooling? Are all of your peripheral devices clean and operational? Once you have run this checklist, you can tell the help desk technician that you have looked into all of these things already and found no solutions yourself.

All of that being said, *please* don't be frustrated when a technician checks these things again. It is important that the technician also runs through this checklist to ensure they have looked at all the possible problems before performing a deep dive into the system to determine where the problem is. There is nothing more frustrating for everyone involved than spending hours trying to fix a network problem and discovering the cable was bad and everyone assumed someone else had already checked that possibility.

ANTIVIRUS

No discussion of problem solving and help desk issues can be complete without an overview of virus and antivirus systems. It is an unfortunate reality of the world that viruses exist. Malicious pieces of software can be designed to be nothing more than an annoyance (advertising pop-ups) or can be designed to lock you out of your computer, encrypt all of your files, and demand payment for the decryption process (ransomware). How do these viruses infect a system? And what can be done to prevent it?

Viruses have a myriad of ways to enter a computer. A computer user can actively infect a machine by opening a piece of software they were either sent or downloaded that unknowingly contains malicious code. Some viruses, known as worms, are programmed to seek out open systems on a network (using standard network learning protocols like a switch would) and then hop from machine to machine without any human interaction required. Others are injected onto a machine through application exploits on various websites. They are hidden in the website code, and once read into a browser for viewing they are able to enter the computer.

So, as you can see, it is entirely possible to get a virus on your machine without *any* action on your part. But it is important to be aware of these attack vectors so that you can attempt to minimize them. Make sure that all software that you download is from a reputable source. Maintain antivirus programs that are designed to proactively keep malicious code from accessing your system. And perform routine antivirus sweeps of your machine to ensure that if anything has gotten in, it is cleaned and removed as quickly as possible.

Any truly good antivirus software is made up of three parts: the updater, which maintains all of the latest virus definitions so that new malicious code can be identified; the firewall, which prevents known malicious code from entering and infecting your machine; and the cleaner, which sweeps all areas of your hard drive for malicious or unexpected code and removes it. Antivirus is one of the few essential pieces of software that all computers require.

KEY IDEAS IN THIS CHAPTER

- Most help desks are set up along the same lines. There is an entry level at which you provide detailed information about what the problem is. Some systems are designed to take this information and find the appropriate technician to respond. Others have a generalist technician who will attempt to fix the problem first before escalating to a specialist.
- Within many school districts, this tiered system is broken down along the generalist/specialist framework, but it is also broken down geographically as well.
- The ticketing system is there so that if more than one technician is required to work on a given problem there are notes on what has already been done, as well as to ensure that work is completed to your satisfaction.
- Service level agreements determine the amount of time it will take to respond to a given issue.
- Rebooting the machine causes the memory buffer to be released. It empties the active programs out of the processor and resets the OS to the initial startup configuration.
- If a reboot does not fix your immediate issue, then physical problems should be the next thing that you look into.
- Viruses have a myriad of ways to enter a computer.
- Any truly good antivirus software is made up of three parts: the updater, the firewall, and the cleaner.

IV

How Do We Build the Box?

TEN

Project Management and Methodologies

Project management has existed throughout human history in some form or another. Undoubtedly, the seven wonders of the world were planned and managed during the building process. However, since the second half of the last century it has been distilled and defined into various complex and different approaches.

At its core, project management is the process for overseeing all of the various aspects of a specific project to ensure the end goal is met on time, on budget, and the desired goal has been achieved. These various processes have been defined and refined by the IT industry since the mid-1950s, and it is here that the most practical world application can be derived from a study of information technology.

Project management, as it is currently practiced, began to take form through the efforts of various civil engineering, US Defense divisions, and early computer programmers. In an effort to control process approach and gain a better understanding of the various aspects and influences of a project, the early methodology was evolved to include timeline and cost estimates. PERT (program evaluation and review technique) and CPM (critical path method) were devised to better determine scheduling approaches when building an order of operations for a project.

IT began to develop its own project management approach when software development began in earnest. The various aspects of bringing a piece of software into fruition can be very time sensitive and labor intensive. It is important, since there is no physical product to see take shape, that you understand what is being coded at all times. There have been a number of approaches that have been defined over the years, and we will explore them now. We will also look at how these approaches have been, or could be, applied to an educational setting.

WATERFALL

Waterfall is the traditional approach to software development project management. It is broken down into very distinct phases and is still the most used approach for software creation. It begins with defining the requirements of the project. This initial phase is very extensive and requires inputs from any and all possible end user types. This is where the goal of the software is being defined. What is it you want the software to do? This will define the entire scope of the project. At the end of this step, the timeline and budget are set for the project.

Next comes the design phase. This is, as the name implies, where all of the various concepts for the look and feel of the software are developed. The function and usability has already been defined by the requirements, so now those aspects must be incorporated into the design in ways that will make sense. After design approval, coding begins. This is broken up into modular pieces, as we discussed in the software chapter. The reasons why this makes sense for how the software is presented to the processor and memory have already been reviewed.

The practical reasons for why this makes sense in creation should be obvious. To enable a schedule that is not overly long, it is necessary to have multiple people working on the various pieces of the software. This does present the problem, though, of having the code exist in pieces and not as a single usable piece of software. Thus, the next phase is integration. This is where the various code fragments are brought together under the overall application framework that will run the software and call the various functions and code portions when needed.

Until now, the software has been built in a vacuum. While the individual functions themselves can be tested to ensure they work while being coded, there is no way to know if there will be problems when the entire project is brought together. Therefore, after integration we enter the testing/debugging phase. This is when all of the original requirements are tested to ensure that the software can meet all of the needs as requested.

Additionally, any bugs in the code that cause the software to function improperly are found and rectified.

This can be a very long phase and can sometimes last even longer than the coding phase. Ensuring that all requirements have been met and there are no obvious problems with the software functionality requires repetitive testing of the same functions. Once this phase has been deemed complete, installation can occur. This is when the software is released to the intended end users. And with that we move into the final phase, maintenance.

Maintenance is the last and longest phase in this process. After release of the finished software, problems will inevitably be discovered. These will need to be fixed. Updates to hardware may impact the software's functionality, so patches will need to be released. And as long as the software is in active use, some level of technical support must be maintained. That is why this phase is by far the longest.

There are many things about the waterfall approach that make it a desirable method for performing a series of tasks. Most especially, it can be useful to determine all of the required end goals before embarking on any other aspects of the project. From an educational perspective, this is the equivalent of determining a program of studies. The key aspect to remember is that the requirements are *gathered* by the design team from the needs of the end users. They are not handed down to the end users by the design team and told "this is what you need." It is a collaborative process.

The biggest downfall to this approach should be self-evident. There are four full phases, and a large portion of the entire work budget and schedule is expended before any testable product has been developed. This can be dangerous if the requirements have not been properly explained or understood. This can, and has been known to, lead to a complete need to redesign the product from the ground up. This will result in cost and timeline overruns, and can sink a project if not properly managed.

AGILE

Agile development is an attempt to combat the downfalls of the linear waterfall method. As the name implies, agile is a rapid development process that uses continuous delivery as a method of refining the product over a series of iterations. The project does not start with extensive and detailed requirements gathering. It begins with the idea of what the end product should be. Then, an initial rough design and build of a portion of the product is produced.

Agile requires near constant communication between the project team and the end user. With each iteration, called sprints, small pieces of the end product are delivered and pieced together. This way immediate feedback and testing can be built into the process. While this seems more chaotic and haphazard, it is in fact a highly disciplined approach to project management. It requires the discipline to not overreach with each sprint.

Again, it is important to ensure that costs and deliverable dates are not exceeded and that each sprint focus only on one new aspect of project definition. Once this has been completed, the next part is begun. Thus the project is brought into shape slowly and methodically over time, not all at once as with the waterfall integration phase. All along communication and feedback from the end user help to ensure that the end goal of the project will still be met and no surprises or misinterpretations can get too far before they are discovered.

The constant feedback and communication model is one that can be applied to many different aspects of education. It is, in fact, a method that can be utilized when integrating technology into different areas and aspects of curriculum development. The teachers and students who will be using the technology to better their skill set should be an integral part of understanding how IT can be better utilized in the educational world.

LEAN

The primary principle behind lean project management is a reduction of waste. Lean focuses on understanding the tasks required for executing a given project, who will be executing the tasks, and reducing the waste that can be generated by focusing on deadlines. By understanding the order in which tasks must be accomplished and who is focused on accomplishing the tasks, the handoff of finished tasks to the next person in line can be streamlined.

This approach does not depend on deadlines to drive work; it depends on the completion of tasks as efficiently and error free as possible. This method, like agile, also requires near constant communication throughout the process: communication among the project team to ensure task handoff is occurring properly, as well as communication with the end users to ensure the tasks are coming together to form the proper end product. The downfall here, of course, is in focusing too much on eliminating waste. It is possible to become too focused on finding ways to cut back and not leaving your team or your project with enough resources to complete the project.

This method can have the most personal application and real-world impact. Not only is it important to be looking for ways to streamline any given project, whether it is in the world of education or elsewhere, but it is also imperative to find ways to reduce waste overall in life. Finding ways to more efficiently flow from task to task, even if the task is relaxation, completing that task, and being able to move on to something else in the knowledge that there is nothing lingering and no need to worry about something that wasn't finished can be life changing.

METHODOLOGIES

While project management approaches have been defined and refined by IT, so too have methodologies for the daily operation of an IT infrastructure. These methodologies can also give insight into ways to structure and perform operational efficiencies in other areas. These frameworks were developed to take advantage of the leverage that technology can give us in performing routine tasks, as well as the capacity enhancements created by technology.

ITIL

The IT infrastructure library (ITIL) is a growing set of methodologies that uses process modeling to define a set of operational management and controls. ITIL covers service support and delivery, infrastructure management, security management, application management, asset management, and implementation management. This is a *very* extensive set of models and regulations for the proper operation of an IT department. The key is the model-based method.

By understanding the various models of how processes should optimally flow, it is possible to understand where the breakdowns in a given process are occurring. It is also possible to manipulate the various models to better understand how operational processes overlap and depend upon each other. There can be many moving parts to operating an IT department, and it is easy to fall into the habit of seeing them in isolation from each other, as a series of tasks and obligations that need to be completed before something else can be done. So it is important to understand how all of the pieces can, and should, interact.

COBIT

Control objectives for information and related technology (COBIT) is a series of governance and oversight tools specifically designed to align IT with business objectives. This is achieved by defining business goals and objectives and linking various IT goals. These goals have metrics established and timelines for determining their maturity and accomplishments. Where ITIL uses modeling to determine optimal process methods, COBIT uses the goal definitions to determine the process methods.

This methodology is easier to adapt for educational purposes. By aligning individual classroom goals and guidelines with the established curriculum, the achievement of those goals can be more easily measured and managed. It is also a methodology by which you can more easily establish which goals are *not* easily achieved or aligned. This review method, where the higher level goals are tested against the abilities of the lower level processes to achieve them, can be a critical piece in curriculum evolution.

There are many more project management and methodology frameworks that can be studied. This small window into these practices has hopefully helped to show how their various applications can be adapted from the IT realm to any larger organizational needs.

KEY IDEAS IN THIS CHAPTER

- Project management is the process for overseeing all of the various aspects of a specific project to ensure the end goal is met on time, on budget, with the desired goal achieved.
- IT began to develop its own project management approach when software development began in earnest.
- Waterfall is the traditional approach to software development project management.

- Agile development is an attempt to combat the downfalls of the linear waterfall method through a rapid development process that uses continuous delivery as a method of refining the product over a series of iterations.
- The primary principle behind lean project management is a reduction of waste by focusing on understanding the tasks required for executing a given project, who will be executing the tasks, and reducing the waste that can be generated by focusing on deadlines.
- The IT infrastructure library (ITIL) is a growing set of methodologies that uses process modeling to define a set of operational management and controls.
- Control objectives for information and related technology (COBIT) is a series of governance and oversight tools specifically designed to align IT with business objectives.

ELEVEN
Process Improvement

As we were just discussing in the last chapter, IT is highly focused on finding methodologies for making work more efficient. Process improvement is an area in which IT is always striving to find the better way to do things. This focus is both internal to the IT department and to the organization as a whole. We know that IT has insight into many different areas, and IT uses that insight to make the organization better.

DEFINING PROCESSES

The first step in fixing any problem is defining the problem. We have already spent a considerable amount of time defining and discussing the various metrics by which we define the IT world. Many of these same metrics can be modified slightly to better understand the organization and the problems that the organization needs to overcome. Let's take a deeper look at the process by which that can be done.

Availability is a key metric, and one that can be translated to many different process areas. In the educational arena, availability can be defined many different ways. At the university level, are certain classes and study areas available to the students? At the K–12 level, what school programs and additional services are available? These, and many other areas, are important to understand and measure so that the problems within them can be defined.

Performance is another key metric that can be translated to find process improvement. Measuring this at the educational level is, obviously, already taking place with the students. There is an ongoing debate about how teachers and administrators have their performance monitored and measured. Again, this book is not designed to pick sides but to show possibilities. By finding ways to measure performance at all levels, it is possible to find where things can be improved.

Change metrics are essential to understanding whether process improvement is working or not. Measuring the starting position of any process change and then the result after the process change is the only way to plot a true understanding of process improvement. Since process improvement should be an ongoing effort, change metrics over time become one of the most essential things to track and measure.

Once the process of defining what will be measured is completed, mapping the process can begin. Start by defining all of the steps involved in performing the process. This may seem tedious, but it is essential to finding where the improvements can happen. This mapping process allows the metrics that have been established to translate to actual steps within the process. Then, the process improvement methodology can be applied.

FINDING THE GAPS

Discovering and determining the areas where processes can be improved is sometimes very easy, but gets harder as processes become more refined. Most times, when starting this process, it is because of an already known area that needs improvement. In fact, it is easiest to start with these because even without proper measurements improvements can be felt once they have been made.

However, before changes are made it is imperative that metrics be established for the process and tracked over a period of time. This period does not need to be excessively long, but it does need to be long enough to ensure there is a solid understanding of what the preimprovement reality of the process is. There is no way to ensure improvement is true and will last without this baseline information.

Once the gaps have been found, it is important to prioritize the order in which they are adjusted and improved. Something that is not often thought of, but needs to be remembered as an essential part of this process, is that making changes in one area of a process can sometimes have an unexpected effect on other areas of the process. So, while measuring the specific changes that are being made, it is important to also be aware of the entire process as a whole during the gap analysis and improvement phase.

Also, it is important that the prioritizing of the gaps ensures that while working on improving one or two specific problem areas no other *new* problems are created. Fixing the speed in which something is accomplished is not helpful if the end result is worse than when the fix was applied. Slow and correct is always better than fast and wrong. So, when prioritizing, keep the ultimate outcome of the process in mind to ensure the integrity of the process goal is maintained.

PROCESS IMPROVEMENT WITHIN IT

IT performs process improvement on an ongoing basis. The cycle of metric definition, process mapping, metric tracking, and improvement implementation is ongoing. As improvements are made, the metrics are reviewed to ensure they are still tracking the proper information. The process map is reviewed and updated as efficiencies are implemented. But how are process improvements defined?

Process improvement must be able to answer one of these questions about the process itself. What value is it providing to the organization? How can that be enhanced? What specific issue is this process meant to address? How can this be streamlined? Does this process create more work, or reduce individual work levels? If it is busy work, how can that be reduced or eliminated?

While this seems like a simple set of questions, the answers will be large and complex. Most processes and tasks that IT undertakes started as an essential function of the organization. But Moore's law shows that over time the hardware (and software) that perform the tasks will become more efficient and capable. So it is imperative that the applications of the tasks themselves be reviewed and updated along with the systems themselves.

So process improvement within IT is an ongoing cycle and one that can be replicated across the organization. What is the task or process that needs to be accomplished? How is it currently accomplished, or what is the first concept for achieving it if it is new? Implement and measure the successfulness of the process. Map the process and find where the inefficiencies are by assessing it against the questions above. Improve the process based on the results found, and start the cycle over again.

PROCESS IMPROVEMENT IN EDUCATION

As the focus of this book is the IT world, let's take some time to look into how IT has already begun to find process improvement within the educational sphere and where there is still room for improvement. IT has already become integrated into the daily lives and activities of teachers at all levels. Administration of a school is accomplished through the use of various technologies and custom designed software. And disruptive technologies, like tablets and educational website development, have already begun to find the gaps in the processes of individual learning styles and fill them.

The basic principle of process improvement through the application of IT resources is to answer the question: "How can this be done better?" While the answer is not always one that can be achieved through the use of technology, more often than not IT can help to find ways to make things faster and require fewer physical resources to achieve the goal. By applying the methodology we have already established for this process, we can and will find ways to make improvements.

Let us walk through an example of how this can be done. Let's look at how books, a physical resource, have begun to make the transition to the digital world. The initial process is very straightforward. Each student as required to have a textbook for the subject they are studying. The process for making this happen is also a fairly straightforward one. Books are purchased based on the number of students who will be attending the class. The books are then distributed to the students.

Right away, there are several gaps that can be found without even putting metrics into place. What if the cost of the books is too much for the budget allowance for the class? What if the number of books available is limited and not enough can be purchased for every student? What is the distribution process for ensuring that all students receive a book? We haven't even gotten halfway through the cycle yet, and already gaps can be identified.

The metrics for this process are also fairly straightforward and can help to easily define the needs for improvement. What is the cost for the books? What is the timeline for receiving them? How many books will be required each session? How much additional inventory of the books will be required to ensure any new students who join late will have access to them? Once these metrics are established, they can be tested and verified over the span of one or two class cycles.

Where can IT improve this process? Again, the metrics and gaps help to define these improvements. While still maintaining the physical assets, IT can be utilized to track inventory and delivery availability. IT can be leveraged to estimate the need for a refresh cycle based on current and future enrollments. By moving from physical assets to virtual books, IT can completely take over the cycle. The costs will lower due to no longer needing to purchase physical books. The timeline for delivery will shrink to zero since a digital asset can be deployed instantaneously. And then the metrics are once again reapplied, and the process is evaluated again for improvement.

CONTINUAL PROCESS IMPROVEMENT

The cycle of improvement doesn't end. Even once a particular process has been evolved to its most optimal solution, other processes will need work and improvement. And as we have seen with some of the amazing leaps in technology, disruptive new advancements can dramatically change the way processes are performed. There can be no "set it and forget it" if the goal is to always find the better, more efficient way.

Therefore, it is important to review all processes on an ongoing basis. This concept of being never satisfied is what has driven every innovation in IT. It is *never* the goal to make changes just to be able to say changes were made. While the cycle is continuous, the operative word is *improvement*. Any changes must be made with that goal, improvement, always in mind.

KEY IDEAS IN THIS CHAPTER

- Process improvement is an area in which IT is always striving to find the better way to do things.
- By finding ways to measure performance at all levels, it is possible to find where things can be improved.

- Discovering and determining the areas where processes can be improved is sometimes very easy, but gets harder as processes become more refined.
- Process improvement must be able to answer one of these questions about the process itself. What value is it providing to the organization? How can that be enhanced? What specific issue is this process meant to address? How can this be streamlined? Does this process create more work, or reduce individual work levels? If it is busy work, how can that be reduced or eliminated?
- The basic principle of process improvement through the application of IT resources is to answer the question: "How can this be done better?"
- The cycle of improvement doesn't end. There can be no "set it and forget it" if the goal is to always find the better, more efficient way.

TWELVE

Risk Management

There are several professions where risk is the daily operating paradigm. Firefighters, soldiers, and policemen come to mind. IT professionals, however, abhor risk. It is one of the factors that IT focuses the most on planning for and actively preventing. There are all kinds of IT-related risks that face the organization, and while some are obvious there are many that most non-IT folks don't even think about.

IT RISKS

The first, and most widely understood, are hardware and software failures. We have already covered the various aspects of what makes up hardware and software, so here we will talk about the various ways they can break down. Hardware, as we have discussed, is simply that—moving parts and pieces that will wear down with age. Computers are run via electricity, and basic physics tells us that when running electricity through a wire some of the energy is lost to heat.

This is why computers have fans and require space around them for air flow. The larger the computer, the larger the size and number of fans needed. The reason that a server is so loud when turned on is because there are so many fans to help pull the heat off of the motherboard. Additionally, processors (where the most electrical work is being done) have heat sinks installed above them to help pull the heat directly off the chip and dissipate it out into the chassis of the computer.

Hard drives also generate heat, but the biggest failure area for them is if the discs or read/write arms are damaged. Much like a record player, the inside of a hard drive is made up of several spinning discs and an arm that reads and writes to and from the discs. These can be damaged, causing physical inability to access the data. The movement to solid state hard drives, hard drives without moving parts, has begun to mitigate against this risk, but the bulk of hard drives still in use are of the spinning disc variety.

Other physical parts of the computer can, and do, go bad. The motherboard develops a short and can no longer carry electrical current properly. The power unit burns out and can no longer power the device. The cable that carries the network traffic (usually made up of copper wire) burns out or develops a break, which causes a network disruption. The list is almost endless when it comes to the possibility of hardware failure.

Software failure can be even more frustrating to deal with. All code, no matter how meticulously tested, has flaws. Sometimes those flaws are only found in certain physical environments. Sometimes they only surface once, cause a disruption that can't be fixed, and upon a reboot are never seen again. And sometimes they can cause a cascading series of failures that corrupts the entire data retrieval system and crashes the entire machine.

These errors, hardware and software, can happen at any time. There are ways to anticipate them, and we will cover that later in this chapter. The errors that are hardest to anticipate, though, are human errors. IT and the technology it manages are only as good as the people who are using the technology. Most human errors only impact the individual who made the mistake, such as having a file overwritten or closing an application without saving.

Where the human error problem is escalated is when it occurs at the enterprise level of the IT systems. Deleting your own saved file is frustrating. Deleting a saved file that is accessed by the entire organization can be devastating. Many of the risk management tools that have been developed by IT have been in response to human error and the need to recover from some kind of human input mistake.

Viruses and malicious software are things that we touched on in the help desk chapter, and we will cover them again in the chapter on security. We should note here, though, that this is an overlapping area of responsibility as software attacks are a risk that need to be planned for and mitigated against. Most virus attacks, while malicious, are not intended to be criminal actions.

Criminal attacks are also a risk area that IT must guard against. IT is the gatekeeper to an extensive amount of private and confidential information. Whether that be the intellectual property of the organization or the personal information of those who are employed there, all of it resides within a computer somewhere. And that makes the safeguarding of that data the responsibility of IT.

Attacks, unfortunately, can come from both without and within the organization. And while that is not something anyone wants to have to think about, it is something that IT needs to be prepared for. Security is an essential part of risk management and has so many parts to it that we will explore it as its own topic soon.

The final area of risk that IT must prepare for is natural disasters. While we tend to think of the Internet and the Cloud as nebulous things that exist separate from the physical world, the truth of the matter is *all* data is data that lives within a physical machine somewhere. Even *the Cloud* is simply a set of servers that are managed somewhere else, by IT professionals who are not part of the organization.

If the organization chooses to maintain its data in the Cloud, then the provider chosen should be able to explain what the contingencies are for natural disasters in their server farm location. If the organization chooses to maintain its data locally, those questions have to be answered by the internal IT team. Depending on where the organization chooses to house its systems, there are various natural disasters that must be planned for whether they be earthquakes, tornadoes, hurricanes, or something else.

RISK MITIGATION

Now that we have explored the various types of risks that IT must contend with, let's look at the tools and methods IT has developed for combating risk. The important thing to remember is there is no way to ensure 100 percent risk aversion. Something will go wrong. Thank you, Mr. Murphy. The only thing that can be controlled is how you are prepared for whatever that possible wrong thing is.

The first step in any determination of risk mitigation is reviewing the legal requirements of the organization. Without any knowledge of what the organization does, it is already possible to list several legal responsibilities that must be taken into account. Privacy of the personal information of all employees is at the top of the list. Ensuring that any and all financial data generated by the organization is properly secured is essential. All intellectual property belonging to the organization must be categorized and secured as well.

In the educational realm, FERPA (Family Educational Rights and Privacy Act) and HIPPA (Health Insurance Portability and Accountability Act) are at the top of the list of regulations that must be understood and followed. COPPA (Children's Online Privacy Protection Act), while outdated, is also a key legal regulation that must be reviewed. All of these, and many more, will be updated and changed as technology advances. So it is important to maintain a solid legal understanding of IT requirements in education.

Once the legal requirements are understood, it is important to move on to an assessment of all of the various risk types listed above. What is the current state of the hardware throughout the organization? What are the software release and update levels? What kind of training has been developed and implemented to help everyone understand how to interface with the technology? What are the antivirus and security procedures for IT? Where is the infrastructure housed, and what needs to be done to ensure its stability and safety?

Hardware replacement should be an ongoing cycle. The risks of older and out-of-warranty hardware have already been established, and the easiest way to mitigate those risks is to replace the hardware. Most hardware can function properly for roughly five years. Within the five- to seven-year window, risk grows exponentially. Not only might hardware die out, but also newer software might not be able to function on such old equipment. And while it can be difficult to think of anything that is only five years old as old equipment, the Moore's law cycle will have seen at least two jumps in physical technology capabilities during that time.

Software updates should also occur on an ongoing cycle. Software manufacturers routinely put out updates to their software, as long as they are actively maintaining that software. It is important to note that *all* software manufacturers do institute an end of life cycle for their software. Once a sufficient number of advanced versions of the software have been released, older versions will no longer be maintained by the manufacturer. This should be taken into account when determining an update and upgrade cycle.

Human error, while hard to predict, can best be mitigated through training and system usage reviews. When problems do occur, those are good times to review what was done incorrectly and how to keep it from happening again. Ongoing training in the use of various IT systems should be undertaken by everyone to help keep this from happening. When something does inevitably happen, backups and disaster recovery systems are an important aspect of repairing the damage.

We've already touched on, and will again review soon, the importance of antivirus protection. It can't be said enough, however, so take this as a reminder to ensure your antivirus is up to date and has swept your computer recently. Since we will be discussing security in depth in the next chapter, we will move past criminal attack mitigation and look at natural disaster planning.

Physical placement of all IT systems must be secure and weather-proof. While this seems as though it should go without saying, computers shouldn't be kept out in the open air. Personal computers, laptops, smartphones, and tablets are capable of resisting a limited amount of certain weather conditions but it is not recommended. Infrastructure equipment (servers, networking equipment, phone systems, etc.) should *always* be housed in a secure indoor environment.

The choice of physical location may be dictated by the location of the organization, but there are data center locations (buildings designed specifically for the housing of IT infrastructure equipment) that can be utilized to ensure better physical security and stability. Physical and geographic redundancy is another way to mitigate natural disasters. Keeping a copy of the organization's infrastructure in a separate physical location that can be used if the primary location goes down can greatly reduce the impact of natural disaster risks.

DISASTER RECOVERY AND CONTINUITY

Even with all of the awareness of possible risks, and the goals and plans in place to keep things from happening, disasters do occur. How IT prepares for and responds to them is an essential function of IT. It is not possible to simply cross your fingers, wish on a star, and hope that bad things won't happen. IT has to assume the worst may one day occur and take steps to ensure the organization can continue to operate even amid these disasters.

The first step in any disaster recovery plan is to ensure that backups of the infrastructure are occurring on a routine basis. The easiest problem to recover from is one where data is lost, or a computer has crashed, and all that is required is a restore from the most recent backup. Backup plans should also include a physical redundancy, as the system that performs the backup is not immune from the same risks as all of the other systems.

It is important to note here, as an aside, that personal backups are just as important as backups performed by the IT department of the organization. Whether it is your personal machine, or the machine that has been designated for you by the organization, data backups should happen on a regular basis. Most IT departments do *not* back up the individual machines within an organization. They are concerned with the infrastructure that runs the organization. And they leave individual-level machine backups to the users of those machines.

The next step in the disaster recovery process is to determine what the essential systems are for the organization to function at a minimal level of IT availability. These systems, whatever they are, must have some form of redundant capability. And when planning for the worst possible scenario (physical loss of all equipment) this means these systems must have some ability to be duplicated. This can take the form of a second set of hardware housed in a separate location, or a virtual backup kept in the Cloud.

These essential system redundancies must be kept as up to date as possible. How up to date will be determined by how essential the function and data on them is. For some, this means the ability to restore at least to yesterday's reality with only the possible loss of a day's work. For the stock exchange, this means real-time backups that allow a restoration with only a second's worth of information lost.

This is the continuity aspect of risk mitigation: What is the lowest level of IT required to ensure the organization can continue if *everything* is lost in a disaster? Once that has been determined, the order in which the infrastructure needs to be rebuilt and recovered from backups must be established. This will mean that some loss of work and data occurs, but nothing that will cause the organization to completely shut down.

Disaster recovery plans should be in place for organizations of every size. Drills and testing should occur once or twice a year to ensure that all redundant systems are capable of handling the full workload of the organization if necessary. Personnel both within the IT department and the organization as a whole should know what a disaster looks like and what the plans are for ensuring continuity of operations.

KEY IDEAS IN THIS CHAPTER

- Risk is one of the factors that IT focuses the most on planning for and actively preventing.
- The first, and most widely understood, are hardware and software failures.
- Hardware is made up of moving parts and pieces that will wear down with age.
- Software code, no matter how meticulously tested, has flaws and can lead to innumerable types of potential problems.
- The errors that are hardest to anticipate are human errors. IT and the technology it manages are only as good as the people who are using the technology.
- Criminal attacks are also a risk area that IT must guard against.
- Depending on where the organization chooses to house its systems, there are various natural disasters that must be planned for, whether they be earthquakes, tornadoes, hurricanes, or something else.
- The first step in any determination of risk mitigation is reviewing the legal requirements of the organization.
- In the educational realm, FERPA (Family Educational Rights and Privacy Act) and HIPPA (Health Insurance Portability and Accountability Act) are at the top of the list of regulations that must be understood and followed.
- COPPA (Children's Online Privacy Protection Act), while outdated, is also a key legal regulation that must be reviewed.
- Hardware replacement and software updates should be an ongoing cycle.
- Human error, while hard to predict, can best be mitigated through training and system usage reviews.
- IT has to assume the worst may one day occur and take steps to ensure the organization can continue to operate even amid these disasters.
- The first step in any disaster recovery plan is to ensure that backups of the infrastructure are occurring on a routine basis.

- Disaster recovery plans should be in place for organizations of every size. Drills and testing should occur once or twice a year to ensure that all redundant systems are capable of handling the full workload of the organization if necessary.

THIRTEEN
Vendor Management

No matter how good, efficient, and hardworking an IT team is, there will always be a need to utilize external vendors. Whether that be simply for purchasing equipment, for individual project support, or for outsourcing entire portions of IT support, vendors, and how they are managed, are an essential aspect of the IT world. *How* an IT department works with these vendors is the key to these relationship successes or failures.

TRUSTED ADVISORS

Here is the bottom line. For a vendor to work best for an IT department (and that is the *key* in this relationship, the vendor must work *for* the IT department, not the other way around), the vendor has to be a "trusted advisor." There is a vast difference between an advisor and an order taker. Order takers call once a month or quarter. They ask how things are going with current hardware, talk up their current specials, and then go away.

Order takers answer when they are called, but all they are doing is filling out an invoice and sending what the IT team says they need. But what if they don't know what they need? What if what the IT team thinks they want isn't right? That's why trusted advisors are the key. A trusted advisor takes the time to get to know the environment, the needs, the plans, and what other vendors are in the mix.

This is a key behavior in essential vendors. None of them act as though the other vendors an IT department is working with are the enemy. The IT department can put two or three of these vendors together on one call and have a very good conversation about what the IT team's needs are, where the pitfalls are, and what they should really be looking to do. This is not a sales pitch where the IT department says, "I think I want this," and the vendor tries to pigeonhole them into a product that the vendor needs to move on their end.

VENDOR SELECTION

Vendor selection can be a frustrating, confusing, and very difficult process. It is easy to fall into the trap of looking at nothing but the bottom line, and while that is an important thing to know about it is *not* the proper driving factor. It can lead to being locked into contracts that are not beneficial to the IT department if they are only focusing on cost. This will leave an IT team working with vendors that they do not trust once the ink is dry and they can see how the vendor really behaves.

It is important to narrow your selection of potential vendors. There are various methods for doing this, but we will be focusing on categorizing vendors based upon the following vendor types: one-stop shop, best of breed, lowest bidder, hybrid, and outsourced. One-stop shop vendors are the most tempting. They can, though, be one of the biggest vendor problem areas. It seems, unfortunately, that anyone who *claims* to be good at everything is not really very good at *anything*.

Contract understanding is so essential with all vendor negotiations, but most especially with one-stop shops. It is important to remember that you can't just accept their contract at face value. While that sounds simple, contracts are not easy and many all-in-one shops have the worst kinds of blanket contracts. Make sure you are looking for clauses that include acceptable service outages and areas of responsibility. This is usually where you find the problem you are facing is not one covered under the contract.

Best of breed is where everyone wants to start, because we all want the best. But the best can be *very* pricy, and so we can't always get that. A more effective way to approach this is to look at the top five vendors in any area. The vendors in slots four and five are going to be pretty good, and they aren't going to demand the same type of pricing as the top vendor. These vendors also tend to be the ones who are most willing to work with you. They have a high-end reputation to uphold, but they aren't so arrogant that they feel they can dictate terms all the time.

Lowest bidder can be decent for startups and in times of budgetary trouble. It can also be good to keep a few discount vendors available *after* you have purchased a few products from them and worked with them on a few projects. Don't discount a vendor because they are lowest bidder, but you should approach them with caution and ensure you aren't sacrificing quality.

Hybrid, as the name implies, is some combination of these. A horrible hybrid is the one-stop/lowest-bidder. This is not a vendor type that is recommended. Outsourced is a vendor type that almost everyone is familiar with. They supply IT services at a lower rate than maintaining an internal staffing position. They generally operate by making a rotating staff of IT professionals available for an IT department's use as needed.

VENDOR RELATIONSHIP MANAGEMENT

Vendor relationships are just that: relationships. They must be maintained and managed. As with all aspects of IT, there are several moving parts to vendor relationships. These will allow for better communication with the vendor, along with mitigation against possible problems before they happen. Contracts are the key to the relationship. The IT department must always have a contract with its outside vendors. Two-way contracts are the best way to ensure that both sides of the relationship know and understand what the responsibilities are for both parties.

Communication must happen on a regular basis. Both sides, the IT team and the vendor, should feel comfortable initiating conversation. These conversations should be substantive and about the needs of the IT department, not what the vendor is currently attempting to sell. Also, any and all vendors should be able to communicate with the IT team at the same time. There will be projects that involve multiple vendors, and it is essential that they feel comfortable talking to each other as well as the IT team.

Find the right representative with the vendor. It is possible to have a good vendor contract and well-laid-out plans, but the person who is the contact with the IT team for the vendor just isn't the right fit for some reason. It is a truth that not all people get along all the time. And there is nothing wrong in talking to a higher authority at the vendor to tell them that the relationship is strained because of the representative assigned to the IT team. Remember, the vendor works for the IT team, and they should ensure that they are properly interacting with the IT team.

Set expectations and ensure the vendor meets them. This is all part of the communication cycle. Make certain the vendor knows what it is the IT team wants and how best to accomplish it. And all of this interaction must be driven by the budget of the process. Budgets should always be a consideration, but they should never be the driving force for a decision. Decisions based solely on budget will hamper the choices available.

VENDOR REVIEW CYCLE

It is important for an IT department to review their vendor group on a regular basis. This is best performed on an annual review cycle. This can also help with new vendors who could call or ask for business. By establishing a vendor review cycle at the same time every year (March into April for example), potential new vendors can send whatever materials they want to be reviewed at that time.

Let's now explore a multiphase approach for vendor review. Contract review is imperative to ensure both sides are staying within the contract. Any issues of note should be dealt with as soon as possible, especially if legal consultation is required. Metrics of "on time" and "on budget" should be established for all projects in the past year. Again, cost and budget are not the *only* factors that should be reviewed, but they are important.

It is also important to know what the upcoming needs and projects for the next year are. Vendors who are trusted advisors should already be aware of what these needs and plans are, and should be working to help define and achieve them. And while it may sound odd when all of these other criteria are based on facts and hard knowledge, it is also important to keep in mind what feels right. Does the IT team work well with the vendor? Are the vendor's personnel pleasant people and easy to get along with? These are important things to note.

All vendor contracts should include a termination timeline. None of them should assume autorenewal. This way, the IT department has to choose to keep them. Some of the decisions to keep or let go of a vendor will be driven by those contract timelines. If the decision is made to bring on a new vendor, then it is important to talk to your outgoing vendor to let them know why you won't be continuing the relationship. It is essential to maintain that communication loop even when you are ending the relationship.

KEY IDEAS IN THIS CHAPTER

- No matter how good, efficient, and hardworking an IT team is, there will always be a need to utilize external vendors.
- For a vendor to work best for an IT department, the vendor has to be a "trusted advisor."
- Vendor selection can be a frustrating, confusing, and very difficult process.
- Contracts are the key to the vendor relationship.
- Remember, the vendor works for the IT team, and they should ensure that they are properly interacting with the IT team.
- It is important for an IT department to review its vendor group on an annual basis.
- All vendor contracts should include a termination timeline. None of them should assume autorenewal.

FOURTEEN

Security

IT security is a vast and extensive topic. It is also one that is imperative for everyone to understand some of the basics. As our world becomes more and more reliant on and run by computer technology, security will continue to rise in importance for everyone. We will review what kinds of security must be maintained, what can and should be secured in various ways, how security is breached, and the ways that security is evolving and advancing with new technologies.

PHYSICAL SECURITY

Physical security is as old as the first fire set to keep the animals away at night. And while the means by which we physically secure IT systems are more advanced, the concept remains the same. Be aware of what could attack. See where the attack could come from. And be prepared to fight off the attack if necessary.

Locked doors and no windows are a must for any IT area, minimizing the ability to physically access the equipment. Some equipment can be physically locked directly on the case, preventing any physical access to the internal parts of the server. Network and server racks can also be locked to minimize access to the equipment itself. And all computer equipment should be kept in a building that itself has doors that can lock and be secured.

Additionally, limiting the physical access to those who are only required to interact with the IT technology is a must. Only IT personnel and other authorized individuals should be allowed into the secure server area. Vendors should have minimal access to only the areas required for them to perform the task they have been hired to do. And no one should be granted access without IT supervision to ensure nothing happens.

NETWORK AND SOFTWARE SECURITY

Network security is a trickier thing to manage, but it follows the same principles. Physical access to network resources should be limited to authorized employees of the organization. Anyone who needs physical connectivity access should be limited via access control lists to ensure they are only accessing the network resources they are allowed to access. Firewalls, redundant and separate Internet connections, and managed switching and routing can all be utilized to ensure that network traffic remains secure.

Access to network resources, once granted access to the network itself, should be managed by secure login and authentication. Logins must be updated, and passwords changed, on a regular basis. Infrastructure systems should have administrative logins that are known only to the IT department. And all login activity should be logged so that any unauthorized activity can be discovered and acted upon.

Software-level security can be managed through both physical and virtual means. Some software is limited to only certain machines on which it can be run. Access to these machines can then be physically monitored and maintained. Some enterprise-level software suites require their own individual account and login to access, even if anyone can see them as available on the network. Individual files can also be encrypted to require password access to read them.

WHAT TO SECURE

As we discussed in the chapter on risk management, there is a list of ways to determine what can and should be secured. At the top of the list should be the physical assets of IT. All of the servers, networking equipment, phone systems, and so forth should be kept both physically and virtually secured. Once that has been accomplished, legally required data must be restricted and secured. This can be both organizational data that falls under various regulations and laws and any data that the organization is contractually obligated to secure by the nature of its work.

Intellectual property is also essential to secure. While this might not fall under a contractual obligation, it is the essential set of information that differentiates the organization. Thus, it must be kept secure and safe. And lastly, any and all work product must be kept secure. Anything that has not already been covered by the contractual obligations or intellectual property security should be, at a minimum, secured from external access.

BREACHING SECURITY

Attacks are, unfortunately, something that all IT departments must prepare for. No one wants to have to think about what might happen in an attack, but it is almost a guarantee that an attack of some kind will take place. IT must be prepared for that and part of that preparation is ensuring that the rest of the organization knows what to look out for.

Physical attacks don't usually look like they do in the movies. When someone is attempting to access the IT infrastructure of an organization through physical means they will rarely (read that as *never*) rappel down from the ceiling on a wire to access the computers. They instead will attempt to gain access through social engineering. Social engineering is the term for physical attempts to circumvent IT security. Generally this takes the form of an individual, or individuals, who are attempting to pose as needing legitimate access to the systems.

Most of us have seen some form of this; we are familiar with the phone calls, letters, or e-mails that attempt this. They reference some organization or authority they claim to be a part of. Then they explain what the problem is they are attempting to solve. And then they close by requesting personal information or access to information for the sake of solving the problem.

It can be very easy to fall for social engineering. While we have become leery of odd sounding e-mails and chain letters, phone calls can still effectively use this method. Those who utilize it are very good at talking a person in circles and being just confusing enough as to why the information is important to hand over. IT staff are trained to ask for certain identifying things, like a ticket or contract number, to ensure that someone asking for information is legitimate.

Network attacks are complicated and highly sophisticated. Generally, the purpose of a network attack is to access data for some kind of malicious use. However, network attacks have also been known to have the motive of solely blocking the capability of an organization to function properly. Network attacks can occur from outside or inside an organization and take multiple forms.

Hacking, a term most people are familiar with, is the general term for anyone attempting to access information for illegal purposes. Internal hacking can occur when an employee purposefully accesses data or systems they are not authorized to for some personal gain. External hacking is any kind of computer-based attack that is designed to find and capture data. Sometimes that data is used for criminal intent. Sometimes the data is simply destroyed or corrupted in an attempt to destroy the organization's ability to function.

Denial-of-service is an Internet-based attack in which a website or online resource is overloaded with requests for response. The overwhelming number of requests forces the resource to go offline as there is no way for it to service that number of requests properly. This attack is solely geared toward halting the proper workflow of the organization. If it is unable to be resolved quickly, it can cause lasting damage to the resource as it overloads its physical capabilities.

Malware and virus attacks are a specialty breed of network attacks. Generally, as we have already discussed, these are meant to be nothing more than a nuisance. They cause systems to act erratically but there is no monetary gain for those who implemented the attack. However, more often lately malware attacks have been instigated with the purpose of forcing or stealing monetary access. Keylogger malware is designed to track all passwords and where they are used to gain access to secure information. Ransomware is designed specifically to disable access to a machine and force the user to pay to have that access restored.

SECURITY MEASURES

The tactics used to secure against all of the various forms of hacking are equally varied and complex. As educators, it is important to know about and be wary of all of the means a malicious attack can utilize. As with any large organization, a school district or university will employ IT-based security procedures to minimize these kinds of attacks. But everyone must note that no matter how much time and effort is put into security, there is no 100 percent foolproof security system.

All organizations will utilize the steps laid out above to secure the various levels (physical, network, and software) and will employ personnel whose responsibility it is to manage security. But it is imperative that everyone is aware of, and wary of, the possibility of malicious attacks. And everyone, no matter their personal level of technological skill, must do what they can to ensure the security of the data they have access to.

EVOLVING SECURITY

While we are still using many of the security measures that were created even before IT, technology has evolved our security possibilities. Password encryption capabilities have become more complex as more difficult algorithms are invoked to build the encoding process. Physical keys have been replaced with IR swipe devices, and even those have begun to give way to biometric security. Soon, individual access to everything will be keyed to our own fingerprints and eye patterns.

No matter the technology involved in security, it is important to know and understand the various ways that security breaches are attempted. The human factor is the most important one to be aware of. Everyone should understand what it is important to secure and how they are a part of the security process, whether it is as simple as not giving anyone their password or as complex as writing the next encryption algorithm to ensure data security.

KEY IDEAS IN THIS CHAPTER

- As our world becomes more and more reliant on and run by computer technology, security will continue to rise in importance for everyone.
- Physical security requires that you be aware of what could attack, see where the attack could come from, and be prepared to fight off the attack if necessary.
- Network and software level security can be managed through both physical and virtual means.
- You should secure the physical assets of IT, legally required data, intellectual property, and any and all work product.

- No one wants to have to think about what might happen in an attack, but it is almost a guarantee that an attack of some kind will take place.
- *Social engineering* is the term for physical attempts to circumvent IT security.
- Network attacks are complicated and highly sophisticated. Generally, the purpose of a network attack is to access data for some kind of malicious use.
- *Hacking,* a term most people are familiar with, is the general term for anyone attempting to access information for illegal purposes.
- As educators, it is important to know about, and be wary of, all of the means a malicious attack can utilize.
- No matter the technology involved in security, it is important to know and understand the various ways that security breaches are attempted. The human factor is the most important one to be aware of. Everyone should understand what is important to secure and how they are a part of the security process.

V

Who Holds the Box?

FIFTEEN
Staffing

IT staffing covers a wide variety of areas. And it is an ever-growing job base that is hungry for new ideas and new innovations. As educators, it is important to know the basics of the various IT jobs to help encourage and guide those who are interested in IT as a career. While many of the titles may sound similar, the breadth and depth of knowledge required for all the various IT job requirements ensures that there will always be differences and challenges specific to each area.

As we have already discussed the help desk in depth, let's start there. It is the entry-level job position for most aspects of IT. It does require a minimal amount of technology knowledge, but it is also the most general area within the IT field. Help desk personnel tend to know a little bit about everything, but not a lot about any one particular area of IT. It is a good position to start in, because it is the one that is closest to the end user.

As IT professionals advance through their career, it is possible for them to forget some of the important lessons learned when working the help desk. Not everyone is a computer specialist. And especially those who are not IT professionals need to be treated with more empathy and understanding, not frustration and contempt. Starting at a help desk allows a growing IT staff member to see where the inevitable problems with the system are and gets them interested in specializing in ways to fix them.

After leaving the help desk, there are several paths to take within the IT world—hardware and systems, networking, security, or application/

web development. Some people are able to generalize in two or three of these areas, but there is simply too much to know to ever specialize in all of them. And it is certainly possible to move from one area to another, as many of the concepts do overlap.

Systems administrators are the first line of defense in the hardware and systems world. They are in charge of managing and maintaining the hardware. Generally, they have some level of responsibility for the software that sits on the hardware, but it is usually limited to ensuring the operating systems are properly functioning. They are not responsible for any of the networking systems, but they can sometimes be responsible for the phone systems. They also usually function as the second tier in the help desk system.

Systems engineers help to implement upgrades and changes to the systems. Where the admins run the equipment, the engineers build and implement the equipment. They are usually in charge of a small group of admins and can also sometimes be in charge of the help desk. They manage project implementations of new system rollouts.

System architects design the systems. They are the most senior members of the systems team and rarely actually interact with the monitoring and maintenance of the current systems. They maintain an awareness of the current system levels, especially as upgrades occur, but they are future focused. Architects are looking at multiyear plans to see what needs to be prepared for and planned in the next three to five years.

On the network side, the job descriptions break down in much the same way. Network administrators oversee the daily use of the network systems. They grant and restrict access as needed. They monitor traffic flow and ensure quality of service is maintained. Network engineers are the builders of the network. It is important for them to know what resources are needed where and what importance is placed on certain types of traffic. Network architects design the network, and much like systems architects, are focused on future needs.

Security roles take a slightly different track. Security administrators are responsible for overseeing all security and safety policies. They will ensure all software vulnerabilities are patched and maintained. They update and manage the antivirus system throughout the organization, and they respond whenever a threat detection has occurred.

Security analysts are not only responsible for developing and designing the security policies of the organization, but they are also responsible for overseeing the training of the staff to ensure all policies are understood and followed. They conduct penetration tests and analysis to ensure the security measures are adequate and maintained. And they are responsible for the physical security of all IT assets as well.

An offshoot of the security world is vulnerability and penetration testing. While this can be a role assumed within the analyst level, there can be sufficient need for this as its own job function. It is certainly a specialty that is sought after. The ability to ensure the security of the systems by attempting to access them through the methods an attacker would use is a valuable tool.

Within the software development world, there are several tracks to take as well. Software developers tend to be limited only by the various programming languages they know. Software development itself is a skill set, but what project a developer can work on will be determined by the need for the programming languages they know. Software testing is, much as it sounds, where the various software requirements are put to the test. Applications are documented and utilized over and over in very specific ways to determine what could break where.

Software development is also varied by the types of software there are to design and develop. Mobile apps are very different from websites. Video games require a team with various skills in both artistic and application design. Operating systems require an intimate understanding of how various hardware will interact with the various applications and how to oversee and manage that process.

While several of the job functions already discussed do have some level of managerial oversight, as with many other jobs, there is delineation between the doers and the managers. IT managers oversee specific areas within IT. They are usually senior-level practitioners of the area they manage. This is also the level at which specialists once again need to start becoming generalists. In order to oversee larger and larger areas of responsibility, it is important to give up the lower-level skills and learn a little bit about everything again.

IT directors and vice presidents generally oversee the entire IT department and all of the various job functions therein. They are no longer managing systems, networks, or applications. They manage budgets and personnel. It is their responsibility to oversee vendor relationship management and overall systems satisfaction. They are usually the interface between upper-level management within the organization and IT.

Above the director/VP tier we enter the C-Suite. Most everyone has heard the term CIO (chief information officer), but there are also CTOs (chief technology officer) and CSOs (chief security officer). Their job functions do have some levels of overlap and are not always separated out at the C-Suite level. Many organizations do not feel the need to single out security for such a high-level position, but if security is a key aspect of the organization this can be a worthwhile position to have.

CIOs are responsible for the overall technological direction of the organization. They propose and manage budgets, approve purchasing and vendor contracts, provide guidance for the future growth and technological upgrades, and manage day-to-day operation of all aspects of the IT department. CTOs differ in that they tend to be more hardware and research based in their efforts. Research and development fall under their purview, as well as any scientific aspects of the organization.

IT JOB PROSPECTS

As with all career paths, IT has its ups and downs. Some jobs that were once the highest paid in the industry are now considered commonplace. New technologies are always changing the various interactions of the job roles and responsibilities discussed here. And though it is true that there are no guarantees in life, IT isn't going anywhere. So while it may be necessary to move between organizations to maintain career growth, there is no doubt that IT offers career stability.

TEACHABLE MOMENTS

Thus far, we have explored all of the various aspects of the IT world. We have talked about how IT can relate to, enhance, and disrupt education. Hopefully you are now more comfortable interacting with those in the IT profession and are prepared to have more meaningful interactions with the generation for whom dealing with technology is like breathing. Now, let us take some time to see what IT has adapted and learned from education.

While the sheer volume of technology in the world means that we are forced to interact with and attempt to understand it, no one is born with an innate technological ability. It is a learned set of skills. How those skills are learned is as important as ensuring that learning is always happening to advance those skills. And as with education as a whole, IT has a series of curricula and teaching tools to make that happen.

APPRENTICE LEARNING

IT adapted and adopted the idea of learning by working alongside someone with more advanced skills early on. This is still practiced in virtually every IT department around the world as team members work together to learn new skills and advance their own personal development. Where IT has truly found the disruption in this methodology is the introduction of virtual presence. Through video and interactive technology, it is now possible to "apprentice" with someone who is a world away physically. IT is continuing to improve upon this method to gain better and faster skill growth.

TRADITIONAL UNIVERSITY LEARNING

IT and IT-related degrees are taught at every major university around the world. These involve theory and practical studies to ensure that students understand and can apply all of the essential concepts of IT methodology. Additionally, the IT community has developed and maintains an extensive set of certification systems that can enhance the learning and skills of any IT professional.

Again, IT has found ways to advance this model through technology-specific institutes and distance learning technologies. IT teaching outreach has stretched down to the lowest elementary grades, starting those students with an early aptitude along the path of IT skill set adoption. And as technology begins to integrate even more into different areas and careers, basic IT understanding will become a requirement for all of those degrees.

CONTINUAL LEARNING

Even after degrees have been obtained, certificates have been awarded, and apprentices have become masters, it is important to maintain a level of IT learning. Moore's law means technology advancement does not stop. Skills that were cutting edge can become unusable within a few short years. In a career that could potentially last forty years, ongoing education is a requirement to stay employable within the IT world.

TEST ENVIRONMENTS

Another teaching tool that IT adopted was the concept of the test environment. Testing environments are separate technical resources that can be used to test new hardware, new software, software updates and upgrades, and new application designs. This separate environment means that none of the essential IT resources for running the organization will be compromised during testing.

Test environments are an essential method for exploring the possibilities of IT. They are designed to be broken, fixed, rebroken, completely taken apart and rebuilt, and redesigned from scratch again. They are modular and useful tools for finding new ways to make IT better. And they are equally important in advancing individual skill sets. Hands-on learning is an important aspect of IT skill building.

The importance of being able to make mistakes cannot be undervalued. Technology has such a hold on our lives that we are afraid of what breaking it could mean. But there is no greater teaching tool for searing a technology fix into the brain than having to survive an emergency of some type. When something is broken and the organization simply *can't* function without it, the way it was fixed and the discovery of what caused it to break will never be forgotten.

But before those kinds of crises can be averted or dealt with, it is important to spend time breaking and fixing nonessential infrastructure. Hence, the test environment. It is the area where the IT professional learns to become desensitized to hardware errors and software crash reports. Like a doctor going through a residency to gain the repetition needed to treat each case with ease and professionalism, the test environment trains an IT professional how not to panic.

KEY IDEAS IN THIS CHAPTER

- As educators, it is important to know the basics of the various IT jobs to help encourage and guide those who are interested in IT as a career.

- Help desk is the entry-level job position for most aspects of IT. It does require a minimal amount of technology knowledge, but it is also the most general area within the IT field.
- IT job prospects branch off into several different areas: hardware and systems, networking, security, or application/web development.
- Systems administrators are in charge of managing and maintaining the hardware.
- Systems engineers help to implement upgrades and changes to the systems.
- System architects design the systems.
- Network administrators oversee the daily use of the network systems.
- Network engineers are the builders of the network.
- Network architects design the network, and much like systems architects, are focused on future needs.
- Security administrators are responsible for overseeing all security and safety policies.
- Security analysts are not only responsible for developing and designing the security policies of the organization, but they are also responsible for overseeing the training of the staff to ensure all policies are understood and followed.
- Software developers tend to be limited only by the various programming languages they know. Software development itself is a skill set, but what project a developer can work on will be determined by the need for the programming languages they know.
- IT managers oversee specific areas within IT. They are usually senior-level practitioners of the area they are given management of.
- IT directors and vice presidents generally oversee the entire IT department and all of the various job functions therein.
- CIOs are responsible for the overall technological direction of the organization.
- CTOs differ in that they tend to be more hardware and research based in their efforts.
- How those skills are learned is as important as ensuring that learning is always happening to advance those skills. And as with education as a whole, IT has a series of curricula and teaching tools to make that happen.

VI

Your Personal Box

SIXTEEN

Living a Digital Life and IT Consumerization

Personal IT use took off with the PC evolution of the 1970s and 1980s. Since that time, our lives have become more and more digitally focused. Once we all had a PC, either at home or at work, it was followed soon thereafter by personal cell phones. Then the Internet exploded and we all had e-mail and websites. Social media took off, and we all began to live interconnected digital lives. Smartphones changed the game again, combining all of our various connection methods into one device.

There is no denying that IT is everywhere. It is so important to understand what this means for our lives now and in the future. Here we will explore the personal side of IT, what that means at home and at work, and what the impact of personal technology in the classroom has been. In the next chapter, we will look ahead at what the next several years has in store for IT both professionally and personally.

PERSONAL DIGITAL EVOLUTION

Most of us didn't have a digital presence until our first e-mail address. This form of communication, that is virtually the default for how we converse now, began our introduction into a digital existence. It was a relatively easy transition. We were used to having phone numbers and home addresses. And in the beginning it wasn't as obtrusive and constant as it is now.

Along with e-mail, early blog sites gave people the opportunity to start presenting their personal thoughts and point of view to the world at large via the Internet. Chat rooms and message boards also became a new communication tool. We discovered that people of all types, when given access to like-minded individuals, wanted to be able to converse and communicate with them.

Social media was born from this desire to interact and connect with people no matter where they are located. The Internet has allowed the world to become a global interactive community, and social media continues to find new ways to present personal thoughts and opinions to the world. And while whether this is good or bad is debatable, there is no denying the impact that social media has had in just the few short years it has existed.

Smartphone technology was the last key piece in creating our digital lives. While most people had some form of digital presence before smartphones, it was their introduction to the masses that truly pushed the digital revolution into overdrive. Once a smartphone or tablet became a part of our everyday life, we never unplugged from the digital world. We are constantly bombarded by e-mail, text messages, updates from our various social media platforms, and sometimes even an actual phone call!

With all of that as our new reality, it is imperative to remember there is no separation anymore between our digital selves and our real life. The person we present to the world via our digital devices and technologies *is* the person the world sees. Even those of us who know each other IRL (in real life) use the digital information presented to us to form thoughts and opinions of each other. And no matter what the website or application promises, once it has been released to the Internet there is no deleting whatever that information is.

As such, while it is never a good idea to live a life of seclusion and paranoia, it is important to regulate and monitor your digital life. Potential employers do utilize social media to learn about job candidates. A lost phone that is connected via apps to every account and system that you own (credit cards, bank account, online pay apps, etc.) means that someone else could potentially have access to all of those pieces of your personal information. Make sure that, if you are utilizing all of those systems, you have sufficient security built into your device.

DISRUPTIVE TECHNOLOGY

By their very definition, disruptive technologies cause upheaval upon their release. Disruptive technologies are the things that come along that no one anticipated that completely change the way something is produced, delivered, or consumed. Fire was a disruptive technology. So was YouTube. And the nature of Moore's law means that disruption will occur faster and faster now.

The latest example of disruptive technology is 3D printer technology. Shortly after this technology burst onto the scene, it was adapted for use in creating artificial limbs cheaply and quickly. It has been adjusted upward in size and is now being tested as a means for creating low cost, quickly constructed housing. Adaptations on the printing materials are being worked on to enable printing of food, again at low costs and with rapid deployment. The limits of what this technology can enable have not even begun to be discussed.

Virtualized environments and advances in 3D visualization technology are poised to be the next big disruptive technology. This might include virtual and enhanced reality glasses that will allow us to experience and interact with the world around us in a much more advanced way, from simple things like having news and interesting facts about the things we are looking at pop up for us if we are interested, to much more advanced things like interactive 3D environments that will allow us to manipulate and maneuver within a completely digital world.

No matter what forms these kinds of disruptive technologies take, the reality is that Moore's law can all but guarantee that we will start to see these occurring on an almost annual basis in the near future. What we will all need to determine for ourselves is just how disruptive to our own personal IT existence we want them to be. Some will embrace every new technology, while others will fight to their last breath to hold on to what they see as the "proper" way things should be.

Change is never easy. And change on such a scale as these technologies have brought us, that can revolutionize (and in some cases entirely replace) ways that we go about our lives on a daily basis, can be the most difficult to understand and embrace. But these kinds of changes can also be the most exciting and rewarding on a personal level. And finding ways to embrace them, and to fold them into our daily lives, will be its own kind of personal technology evolution.

IT CONSUMERIZATION

The IT world used to be full of gatekeepers—those who held the access keys and ensured that technology was handed out only to those who were deemed able to properly use the technology. And the non-IT masses understood that this was the way technology was to be regarded. It was a complex and difficult world that only a select few had the skills to penetrate. The PC boom began to change this, though.

Once it was possible for everyone to not only have a computer in their own home, but also build it themselves, the equation began to change. The Internet boom, which included personal access to e-mail and other technical resources that were usually only controlled and handed out by the IT department, continued this change in the way IT and technology were viewed. If it was possible to find a piece of hardware or software for purchase online, why does the IT department need to be bothered?

As with so many other things when it comes to personal IT, the smartphone pushed everything over the edge. Once smartphones and tablets became available, and they were capable of performing the same kinds of tasks as a work computer, massive imbalances began to appear between corporate IT and personal IT. The corporate world found it hard to keep up with the rapid advancements of the personal IT revolution.

There are many reasons for this. The smartphone came at a tipping point in the Moore's law curve. This meant that rapid jumps in technology began to happen almost annually at that point. Most corporate IT operated on a multiyear roadmap. They looked at potential hardware and software advancement needs on a three to five year plan. Budgets, projects, resources, and timelines were built and developed with those goals in mind.

Technology that outpaces those plans at nearly three times the speed simply can't be competed with. And even when those plans are altered to take advantage of newer and more rapidly advancing technology, the infrastructure of most organizations is simply not robust enough to offer the same kinds of experiences companies that provide technology resources on the personal level can.

The reason that Microsoft, Apple, and Google can offer so many resources at relatively low prices is because of the sheer number of people who are purchasing those products. There are very few organizations that can replicate the amount of resources companies of that size can provide. So the personal experience with technology is now different from what an organization can provide to its employees.

SHADOW IT

This phenomenon has given rise to what IT calls shadow IT. This is when IT resources are sourced, purchased, and implemented without the knowledge of the IT department. This can cause problems when unanticipated systems attempt to gain access to the organization's IT resources. As we've already covered in the risk and security sections, there are massive potential problems in bringing in unknown and nonapproved technology to an organization.

IT has attempted to battle the rise of shadow IT through various security measures. Network access and resource usage can be physically restricted to minimize the addition of nonauthorized technology. Policies prohibiting the use of personal devices can be instituted throughout the organization. But the easiest way to minimize the rise of shadow IT is to consult IT on all potential technology needs. While it may be faster, sometimes, to purchase the equipment without IT authorization, the walls that IT puts up to prevent the usage of those resources cost time and effort on all sides.

PERSONAL IT IN THE CLASSROOM

The use of personal technology in the classroom is an evolving reality. Most university classes look incomplete without a laptop in front of each student these days. Smartphones are present in every student's pocket. Their use can be hard to restrict, especially when there is obviously a precedent for technology use with all those laptops present. However, there is nothing wrong with requesting the common courtesy of no phone calls in a class.

In the lower grades this is a different debate. Younger and younger children now have smartphones and laptops. The resources that are available by allowing the use of those devices is hard to argue against. However, it is also understandable that they can be more of a distraction than an assistance at such a young age. But, on the other hand again, we don't want to hamper the growing skill set of young technology users when their ability to use technology is going to be such an important part of their lives as they get older.

What it is important to prepare for is how these personal devices can make the learning experience such an individualized one. Personal technology, within the educational context, can be used by teachers as a means to bring the interests of the students into the classroom. Utilizing these devices as a teaching tool, or as an enhancement to the resources already available within the classroom, is a skill set that many teachers are already embracing.

This debate will continue and evolve as technology grows. As we will see in the next chapter, the near future of technology shows that IT will become even more integrated into our daily lives. Once IT has become a part of our clothes, is what we use instead of glasses, and is embedded in our very skins, how will we be able to keep it out of the classroom? So, while it is important to teach the skills of learning *without* technological assistance, it won't be too long before the idea of being separated from IT in any way is anachronistic.

KEY IDEAS IN THIS CHAPTER

- Personal IT use took off with the PC evolution of the 1970s and 1980s. Since that time, our lives have become more and more digitally focused.
- With all of the various technologies available as our new reality, it is imperative to remember there is no separation anymore between our digital selves and our real life. The person we present to the world via our digital devices and technologies *is* the person the world sees.
- Disruptive technologies are the things that come along that no one anticipated that completely change the way something is produced, delivered, or consumed.
- The infrastructure of most organizations is simply not robust enough to offer the same kinds of experiences that companies that provide technology resources on the personal level can.
- Shadow IT is when IT resources are sourced, purchased, and implemented without the knowledge of the IT department. This can cause problems when unanticipated systems attempt to gain access to the organization's IT resources.

- The use of personal technology in the classroom is an evolving and hotly debated reality.
- Personal technology, within the educational context, can be used by teachers as a means to bring the interests of the students into the classroom. Utilizing these devices as a teaching tool, or as an enhancement to the resources already available within the classroom, is a skill set that many teachers are already embracing.

SEVENTEEN
Future Watching

There is no doubt that technology is here to stay. And once again, we look to Moore's law to see that rapid advancement will likely continue to increase. PCs begat laptops, which begat tablets and smartphones, which have already begun to beget new forms of technology to interact with. Advances in product development and rapid creation have begun with 3D printing and have already started to evolve. And the world of biotechnology is one that we haven't even begun to discuss within these pages, but it is making major advancements right alongside all of these hardware advancements.

WEARABLE TECHNOLOGY

We are already, as of the writing of this book, seeing an evolution in the way technology is integrated into our very person. Wearable technology is something that science fiction writers have long predicted, and usable and practical wearable tech is now in existence and making rapid advancements. Watches as smart as computers are on the horizon, as well as glasses that act as screens and enhance the very reality we are looking at. And this just scratches the surface.

Medicine is already experimenting with wearables that provide continuous updates on blood pressure, blood oxidation levels, body temperature, blood glucose levels, and more. As the technology continues to miniaturize and be applicable using small patches, these sensors will enable medical care to rapidly asses the vitals of anyone they apply a patch to. The speed at which this will increase the possibility of proper care being administered can sometimes be a matter of life or death.

We already discussed how 3D printing technology has been adapted to create rapid and inexpensive artificial limbs. Wearable technology has also begun the integration of nerve endings into these artificial limbs, thus giving the sensation of feeling in those appendages. The next evolution that is being worked on is the ability to control those limbs with the central nervous system, just like we do with our natural appendages.

Various wearable devices are in development that will facilitate personalized coaching in the sports world. All kinds of metrics on body movement, posture, muscle use, and energy output will be able to be monitored remotely to provide individualized changes and coaching tips. With the addition of the medical technology, these wearables will also be able to monitor and maintain the health of athletes, thus minimizing the number of injuries an athlete could potentially receive.

These are just the beginning of the wearable revolution. Other breakthroughs and advancements are being made in internalizing various technologies. Again, the medical applications are enormous: personalized diet and exercise that is based on our own internal metabolism and the best possible ways to impact it, microsurgery that can be conducted using robots that are ingested by the patient. And much more!

WEARABLE IN THE CLASSROOM

This revolution is not just going to impact the worlds of medicine and sports. The potential applications in the classroom are staggering. Imagine glasses, or even tabletops, with the ability to track eye movement to help gauge the attention levels of the students. Or pulse and respiration monitors to help determine stress levels and to facilitate determining areas of study that are problematic for children.

Treadmills and bicycles are already being used in some classrooms to help with integrating movement into the learning environment. Attaching these to computers would allow students to power their own workstations, while also helping to monitor fitness levels to help with keeping them healthy and active. Again, these things are all just the tip of a very, *very* large evolution in the way we are about to interact with our technology within the next several decades.

THE INTERNET OF THINGS

The Internet of Things refers to the interconnected devices that are now coming to market: appliances that link to the Internet to give updates on internal use, houses that include security and environmental controls that can be accessed via the Internet, refrigerators that can sense the freshness and amounts of the food inside and then send a shopping list update to your smartphone.

This is what we mean by IT becoming integrated into everything that we do and every aspect of life. The technology will be in our furniture, allowing us to conform the chairs and beds to our ergonomic needs based on the sensors in our clothes. IT will be embedded in our pets, allowing us to track their movements and ensure their health along with our own. Predictive algorithms will determine our potential needs and provide access to purchase those items via the smart glass on our windows.

This may sound like science fiction, but this will all be here within the next few years. We have reached a point on the Moore's law curve where advancement is not small when it happens; it is a massive leap forward. Each advancement in the hardware enables more and more application advancement. And those applications require better and more advanced hardware, and the cycle continues.

SMART TRANSPORTATION

Self-driving cars are no longer a question of if they will happen. It is now a question of how quickly they will be adopted by everyone. And self-driving cars aren't the only form of smart transportation that is being adopted. Rapid transit bullet trains are already a reality. The next evolution, the hyperloop, has gone into active development.

Personal transportation will shortly take the form of nonintrusive movement. We will not need to fight transportation methods to get places. We will simply tell our technology where we want to go, and it will determine what the most efficient means is for getting us there. And this doesn't even begin to look at the possible impact that virtual environments could have on travel. The need to physically go places to interact with them will never fully go away, but it will certainly evolve over time.

EVOLUTION OF INFORMATION

The sheer volume of information available to us thanks to the Internet is sometimes beyond comprehension. That volume of data will continue to grow at an exponential rate. Therefore, it is likely that in the next five years, over 50 percent of the things that we as a species will know do not presently exist. Additionally, connection speeds and transfer mediums are evolving at such a rate that near instantaneous data transfer around the globe will be possible within the same timeframe.

This means that as we continue to expand our capacity for knowledge, and the things that we now know, we will be able to broadcast that knowledge worldwide the moment it is learned. This, perhaps more than any other upcoming advancement, will have a truly disruptive impact on education. The collective knowledge of the world is already available to us, but when we are all wearing a computer and can access groundbreaking knowledge at the very instant that it becomes available, how quickly will our capacity for learning evolve?

THE SINGULARITY

All of this technological evolution is converging on what is known as the singularity. This is the moment at which humanity and technology truly merge. Many look at this point as when human consciousness will be convertible to digital content. Others view this as the point when true artificial intelligence has been created. When this may occur is still a matter of intense debate, but the fact remains that technological evolution is no longer science fiction.

It is certainly still a ways away. And we don't really know what it will mean or look like when it does occur. But there is no doubt that over the next several decades we will continue to see the evolution of our personal connection to technology. The future is a very bright and exciting one. And IT will always be there to help shepherd that future along. Just don't forget to reboot.

KEY IDEAS IN THIS CHAPTER

- There is no doubt that technology is here to stay. And once again, we look to Moore's law to see that rapid advancement will likely continue to increase.
- Wearable technology is something that science fiction writers have long predicted, and usable and practical wearable tech is now in existence and making rapid advancements.
- This revolution is not just going to impact the worlds of medicine and sports, as we will also see rapid classroom application and functionality.
- The Internet of Things refers to the interconnected devices that are now coming to market, and what we mean by IT becoming integrated into everything that we do and every aspect of life.
- Personal transportation will shortly take the form of nonintrusive movement. We will simply tell our technology where we want to go, and it will determine what the most efficient means is for getting us there.
- It is likely that in the next five years, over 50 percent of the things that we as a species will know do not presently exist.
- The future is a very bright and exciting one. And IT will always be there to help shepherd that future along.

VII

Closing the Box

EIGHTEEN

Conclusion

Thank you for taking the time to go on this crash course with me. I hope that this book, while there were *a lot* of technical concepts thrown around, was still grounded and understandable enough to be of use to you. It is simply impossible to discuss IT without getting somewhat technical, so thank you for hanging in there during those parts.

Finding ways to make IT accessible to everyone is not always easy, but if the last two chapters of this book have shown anything, it is that it is essential to understand and accept IT throughout our lives. It will not be going away, and it will only become more and more of an essential tool in the fabric of everything that we do. And when the time comes that you will be faced with that inevitability, you don't need to be so scared.

Now that the overview of the facts and facets of IT and how it interacts with the organization and education has been completed, I would like to take this opportunity to voice my personal thoughts on technology in education. This is something that I struggle with for many reasons. And I honestly do see and understand both sides of the argument.

Because my wife is a teacher, I am aware of the distractions inherent in allowing too much technology into the classroom. I know what technology is currently being used and what level of technological knowledge she is required to have to ensure she knows how to use all of the tools effectively. I understand how hard it can be to maintain classroom discipline in the best of times. Adding a personal distraction device (insert your favorite piece of portable technology here) can only make it harder.

I also have school-age children. I see on a daily basis the allure that computers have for them. My oldest, in fact, amazes me at his capacity to be playing an online video game with his friends, whom he is talking to via video chat on a separate device, while he is watching a video explaining how to work through the level they are on, *and* consulting a book that he has about the game. This ability to take in so many forms of input and successfully process them is astonishing to me.

And here is where my IT background really kicks in and starts to fuel the other side of the argument for me. Technology is essential. And on a daily basis, I interact with people who are either intimidated by it or simply cannot understand the leverage that technology can offer. We don't exist in a world where you can get by with a calculator and a piece of paper anymore. Sometimes I wish that we did, but we simply don't.

I know what is coming in the technology world. I know that the advancements in hardware are going to make more capacity and greater performance available at lower prices and in smaller devices. All of that translates to devices that are even better than our current generation of smartphones, and they'll be on our wrists. Or work as a headband. And before we know it, they may simply be implanted at birth. (Yes, this is a ways off, but it just might happen.)

So because of all of that, I simply don't understand why there is such a fight to keep technology out of the classroom. Technology is essential. Understanding how to use it, when to use it, what is appropriate usage, and what isn't are all things that should be taught right beside math and reading. We are already teaching our children how to properly use resources when we teach them how to respect other people's property and how to properly hold a book.

Why does technology need to be any different? We already have the answers to the questions on the test in the book, but we teach children not to use that when the test comes around. They need to know how to learn and maintain that information in their heads. Why shouldn't using the resources that technology affords us be taught the same way? Yes, you can use your tablet or smartphone when working on the assignment as part of the learning process, but you still need to be able to retain the data for the test.

Technology is a tool. Like the typewriter and the encyclopedia before it, the learning environment only got better when those tools were integrated into classroom use. And thus, we need to find ways to integrate technology into the classroom more. I understand there are limits to resources, and we need to ensure that everyone has a fair advantage. I'm certainly not advocating for any kind of technological class system.

But I do think that everyone should be afforded the opportunity to learn and utilize the tool sets that technology affords us. Education needs to stop fighting technology and learn to embrace it, to stop fearing the disruption and learn to use the disruptive capabilities that technology has created. I have started to see this happen already, and I wholeheartedly encourage it to continue.

So now I will get down off of my soap box. Technology is here to stay, and like water finding its level, it will be used to the best of the capacities of those who are using it. Thank you for your time. Now, hit the save button. Make sure this has all been moved to permanent memory. And shut down.

About the Author

Christopher McCay has worked in the IT field for sixteen years, the last six of them as director of information technology for Brailsford and Dunlavey, a national program management firm. When he isn't playing with technology, he is enjoying time with his wife and two sons, playing video games, and spending time with their pets.